# A STRATEGIC VISION
## FOR DEPARTMENT OF ENERGY ENVIRONMENTAL QUALITY RESEARCH AND DEVELOPMENT

Committee on Building a Long-Term Environmental Quality Research and Development Program in the Department of Energy

Board on Radioactive Waste Management

Division on Earth and Life Studies

National Research Council

NATIONAL ACADEMY PRESS

Washington, D.C.

Support for this study was provided by the U.S. Department of Energy under Grant No. DE-FC01-99EW59049. All opinions, findings, conclusions, or recommendations expressed herein are those of the authors and do not necessarily reflect the views of the U.S. Department of Energy.

International Standard Book Number 0-309-07560-2

Additional copies of this report are available from:

National Academy Press
2101 Constitution Avenue, NW
Box 285
Washington, DC  20055
(800) 624-6242
(202) 334-3313
Online at: http://www.nap.edu

# THE NATIONAL ACADEMIES

National Academy of Sciences
National Academy of Engineering
Institute of Medicine
National Research Council

The **National Academy of Sciences** is a private, nonprofit, self-perpetuating society of distinguished scholars engaged in scientific and engineering research, dedicated to the furtherance of science and technology and to their use for the general welfare. Upon the authority of the charter granted to it by the Congress in 1863, the Academy has a mandate that requires it to advise the federal government on scientific and technical matters. Dr. Bruce M. Alberts is president of the National Academy of Sciences.

The **National Academy of Engineering** was established in 1964, under the charter of the National Academy of Sciences, as a parallel organization of outstanding engineers. It is autonomous in its administration and in the selection of its members, sharing with the National Academy of Sciences the responsibility for advising the federal government. The National Academy of Engineering also sponsors engineering programs aimed at meeting national needs, encourages education and research, and recognizes the superior achievements of engineers. Dr. Wm. A. Wulf is president of the National Academy of Engineering.

The **Institute of Medicine** was established in 1970 by the National Academy of Sciences to secure the services of eminent members of appropriate professions in the examination of policy matters pertaining to the health of the public. The Institute acts under the responsibility given to the National Academy of Sciences by its congressional charter to be an adviser to the federal government and, upon its own initiative, to identify issues of medical care, research, and education. Dr. Kenneth I. Shine is president of the Institute of Medicine.

The **National Research Council** was organized by the National Academy of Sciences in 1916 to associate the broad community of science and technology with the Academy's purposes of furthering knowledge and advising the federal government. Functioning in accordance with general policies determined by the Academy, the Council has become the principal operating agency of both the National Academy of Sciences and the National Academy of Engineering in providing services to the government, the public, and the scientific and engineering communities. The Council is administered jointly by both Academies and the Institute of Medicine. Dr. Bruce M. Alberts and Dr. Wm. A. Wulf are chairman and vice chairman, respectively, of the National Research Council.

*iii*

# Preface

The U.S. Department of Energy (DOE) is responsible for addressing a host of environmental problems associated with radioactive, hazardous, and mixed low-level wastes, nuclear materials, spent nuclear fuels, and contaminated lands, waters, and buildings[1] at over a hundred sites throughout the United States. DOE estimates that the nation will spend over $200 billion to remediate, manage, and dispose of these wastes and contaminated media over the next 70 years (DOE, 2000e). Even after many contaminated sites have been "cleaned up" in accordance with applicable regulations, residual risks to human health and the environment will remain at most DOE sites for centuries, if not millennia, and therefore will require some form of long-term stewardship (DOE, 1999a; NRC, 2000a). DOE currently spends approximately $6.7 billion a year on activities to manage and dispose of wastes and contaminated media throughout the DOE complex (see Sidebar 2.1 for description of the DOE complex). These activities are termed DOE's Environmental Quality (EQ) business line.[2]

Approximately 4 percent of DOE's EQ business line budget is spent on research and development (R&D) activities to improve scientific understanding and develop new approaches to address EQ problems. Since 1998, DOE has referred to these activities as its EQ R&D portfolio. The first comprehensive description of the portfolio was published in February 2000 (DOE, 2000b). In compiling this description, DOE recognized that its EQ R&D portfolio "may be under invested to sustain achievement of existing mission objectives beyond the near term, i.e.,

---

[1] The committee refers to these diverse types of waste, spent fuels, nuclear materials, and contaminated media collectively as "DOE wastes and contaminated media" (see Sidebar 1.1).

[2] EQ is one of DOE's four programmatic business lines. The other three programmatic business lines are Energy Resources, National Nuclear Security, and Science (see discussion in Chapter 1). The four programmatic business lines are supported by a corporate management function, which DOE's most recent strategic plan refers to as a fifth business line (DOE, 2000f).

beyond 2006" (DOE, 2000b, p. xiii).[3] This recognition prompted the Under Secretary of DOE to ask the National Academies' National Research Council (NRC) to provide advice on how DOE's EQ R&D portfolio could broaden its current short-term focus to include a more long-term, strategic view.

The committee was asked to address the following four questions, focusing on post-2006 R&D:

1.  In the context of EQ strategic goals and mission objectives, what criteria should be used to evaluate the adequacy of the portfolio?
2.  Using these criteria, what should be the principal elements of the portfolio?
3.  Should the portfolio be designed to address environmental problems outside DOE (e.g., Department of Defense, Russia) that are related to EQ strategic goals?
4.  How to determine the level of future investments in EQ R&D?

These questions differ from many NRC task statements in that they focus on high-level R&D management issues rather than detailed questions about a specific scientific or technical issue. Taken together, the answers to these four questions constitute the committee's views of how DOE's EQ R&D efforts can be made more effective by focusing more explicitly on DOE's most challenging EQ problems, i.e., a "strategic vision for DOE EQ R&D." The task statement also is unusual for the NRC because it asks for advice related to R&D funding levels. In particular, the committee was asked to provide advice on how to determine the level of future EQ R&D investments. It is important to recognize, however, that the committee was not asked to recommend a level of funding, nor to comment on whether the current level is too high or too low.

The task did not call on the committee to conduct a detailed evaluation of the existing EQ R&D portfolio. Such an analysis was conducted last year by DOE's Strategic Laboratory Council (referred to throughout the report as the "adequacy analysis" and summarized in Appendix C). This report complements and builds on the results of the adequacy analysis and also relies strongly on recent analyses of parts of the EQ R&D portfolio that have been carried out by other NRC committees (see annotated bibliography in Appendix F) and other review groups.

This study could not have been completed without the assistance of many individuals and organizations. The committee wishes to thank the many DOE staff members in the Office of Environmental Management; the Office of Civilian Radioactive Waste Management; the Office of Nu-

---

[3] The portfolio's short-term emphasis has been confirmed by two subsequent analyses of DOE's EQ R&D portfolio (DOE, 2000g,h).

clear Energy, Science and Technology; and the Under Secretary's office for their active participation in committee meetings and in responding to requests for information. The committee is especially grateful to Gerald Boyd, who served as DOE's primary contact for this study, and his staff, particularly Mark Gilbertson, Jef Walker, Ker-Chi Chang, and Lana Nichols.

The committee expresses its deep appreciation to everyone who participated in the committee's two-day workshop in August 2000 (see Appendix B). The diverse mix of participants from DOE (headquarters and the sites), other agencies, national laboratories, academia, nongovernmental organizations, and the private sector contributed to lively discussions that provided great insights into the committee's task. The committee is grateful to speakers Jack Gibbons, David Heyman, James Owendoff, and Ivan Itkin, who helped set the stage for the workshop discussions.

This report has been reviewed in draft form by individuals chosen for their diverse perspectives and technical expertise, in accordance with procedures approved by the NRC Report Review Committee. The purpose of this independent review is to provide candid and critical comments that will assist the institution in making the published report as sound as possible and to ensure that the report meets institutional standards for objectivity, evidence, and responsiveness to the study charge. The content of the review comments and draft manuscript remains confidential to protect the integrity of the deliberative process. We wish to thank the following individuals for their participation in the review of this report:

John F. Ahearne, Sigma Xi and Duke University
John Applegate, Indiana University School of Law
Allen G. Croff, Oak Ridge National Laboratory
James Economy, University of Illinois
John Fischer, U.S. Geological Survey (retired)
John C. Fountain, State University of New York at Buffalo
Thomas Leschine, University of Washington
Alexander MacLachlan, E.I. DuPont de Nemours & Company (retired)
John Pendergrass, Environmental Law Institute

Although the reviewers listed above have provided many constructive comments and suggestions, they were not asked to endorse the conclusions or recommendations, nor did they see the final draft of the report before its release. The review of this report was overseen by Michael Kavanaugh (Malcolm Pirnie, Inc.) and Paul Barton (U.S. Geological Survey, retired). Appointed by the NRC, they were responsible for making certain that an independent examination of this report was carried out in accordance with NRC procedures and that all review comments were carefully considered. Responsibility for the final content of this report rests entirely with the authoring committee and the NRC.

Finally, the committee thanks the NRC staff who assisted the committee throughout the study. Latricia Bailey provided very strong administrative support in all phases of the study, especially during committee meetings and in the preparation of the report. Suzanne Pessotto was instrumental in ensuring the success of the committee's summer workshop by doing an exceptional job handling all of the logistical challenges. Susan Mockler provided research support and prepared meeting minutes. Jennifer Nyman, a summer intern with the Board on Radioactive Waste Management, assisted in information gathering activities early in the study. Kevin Crowley, director of the Board on Radioactive Waste Management, provided helpful strategic advice to the committee. Gregory Symmes, the study director, was of invaluable assistance to the committee in preparation for and during the workshop and the other meetings and in turning committee members' writing into a cohesive and effective report.

Gregory R. Choppin
*Chair*

# Contents

# EXECUTIVE SUMMARY

······························································································

The U.S. Department of Energy (DOE) is responsible for a diverse range of radioactive, hazardous, and mixed low-level wastes; nuclear materials; spent nuclear fuels; and contaminated lands, waters, and facilities (hereafter referred to collectively as "DOE wastes and contaminated media"). These wastes and contaminated media present the following general scientific, technical, and social challenges that will endure long into the future (this list of challenges, which was developed by the committee, is discussed more fully in Chapter 2):

- Remediate (i.e., "clean up") DOE sites[1] and facilities that have severe radioactive and hazardous waste contamination from past activities.
- Manage, stabilize, process, and dispose of a legacy of widely varying and often poorly characterized DOE wastes (including spent nuclear fuels and nuclear materials treated as waste) that are potential threats to health, safety, and the environment.
- Provide effective long-term stewardship[2] of DOE sites that have been remediated as well as currently practical, but that have residual risks to health, safety, and the environment.
- Develop, open, and operate unique, first-of-a-kind facilities for the permanent disposal of radioactive spent fuels and high-level wastes—many of which will be hazardous for thousands to hundreds of thousands of years.
- Limit contamination and materials management problems, including the generation of wastes and contaminated media, in ongoing and future DOE operations.

DOE currently spends approximately $6.7 billion a year to address

---

[1] See Sidebar 2.2 for a description of the largest DOE sites.
[2] DOE defines long-term stewardship as "all activities necessary to ensure protection of human health and the environment following completion of cleanup, disposal, or stabilization at a site or a portion of a site."

1

these challenges through the activities of its Offices of Environmental Management (EM) and Civilian Radioactive Waste Management (RW), and some programs within the Offices of Nuclear Energy, Science and Technology (NE) and Fissile Materials Disposition. DOE refers to the activities addressing these challenges collectively as its Environmental Quality (EQ) "business line."[3] Approximately 4 percent of the total EQ business line budget, or $298 million is spent on research and development (R&D) designed to support the EQ business line. DOE refers to these R&D activities collectively as its "EQ R&D portfolio."

The National Academies' National Research Council undertook this study in response to a request from the Under Secretary of Energy to provide strategic advice on how DOE could improve its EQ R&D portfolio. In particular, the committee was asked to address the following four questions, focusing on post-2006 R&D:

1.  In the context of EQ strategic goals and mission objectives, what criteria should be used to evaluate the adequacy of the portfolio?
2.  Using these criteria, what should be the principal elements of the portfolio?
3.  Should the portfolio be designed to address environmental problems outside DOE (e.g., Department of Defense, Russia) that are related to EQ strategic goals?
4.  How to determine the level of future investments in EQ R&D?

## SCOPE OF DOE's EQ MISSION

It is important to discuss the scope of DOE's EQ mission because any consideration of the adequacy of an R&D portfolio requires a clear understanding of the programmatic objectives that these R&D activities are intended to support. Such clarity is a challenge, however, because DOE's use of the term "environmental quality" is a misnomer that creates a great deal of confusion, both within and outside DOE, and because DOE documents reviewed by the committee are not entirely consistent in describing the EQ mission. The committee discusses the following three aspects of this issue: (1) its topical breadth (i.e., whether it includes, or should be broadened to include, environmental issues beyond wastes and contaminated media); (2) its temporal breadth (i.e., whether it should focus more explicitly on longer-term problems); and (3) its national and international breadth (i.e., whether its responsibilities should be extended to problems outside DOE, such as those in other agencies or nations).

---

[3] EQ is one of DOE's four programmatic business lines (the others are Energy Resources, National Nuclear Security, and Science). The programmatic business lines are supported by a corporate management function, which DOE's 2000 strategic plan refers to as a fifth business line.

The EQ R&D portfolio currently has a large number of important R&D opportunities and gaps, especially in areas requiring long-term R&D.[4] The committee believes that it would be inappropriate to consider expanding the topical breadth of the EQ mission until the R&D portfolio adequately addresses these gaps and opportunities. Furthermore, expanding the topical breadth of DOE's EQ mission to include all areas of the environment, such as sustainable development and global environmental protection, would create significant overlap with DOE's other missions (in particular, the Energy Resources and Science missions), as well as the missions of other federal agencies with longstanding environmental responsibilities. **For these and other reasons discussed in Chapter 2, the committee concludes that the EQ mission should continue to focus on problems associated with DOE wastes and contaminated media.** However, this conclusion does not lessen the importance of closely coordinating EQ R&D with related R&D efforts by DOE's other business lines.

One of the most consistent and important findings of recent studies of the EQ R&D portfolio is that it lacks a long-term strategic vision. The committee believes that this is due in part to the rather limited view of long-term EQ responsibilities in DOE's recent strategic plans (especially the 2000 plan). This short-term emphasis has provided a means for making progress on those aspects of the EQ mission for which technologies exist, but has done much less to address DOE's long-term and most challenging EQ problems,[5] such as those associated with the treatment and disposal of high-level radioactive waste and long-term stewardship. This emphasis also may have been misinterpreted by some decision makers to mean that the EQ mission, and particularly its R&D requirements, will be largely completed by 2006 or 2010. This "going out of business within the next decade" view has served to obscure the reality of DOE's long-term EQ responsibilities. **The committee recommends that DOE develop strategic goals and objectives for its EQ business line that explicitly incorporate a more comprehensive, long-term view of its EQ responsibilities.** For example, these goals and objectives should emphasize long-term stewardship and the importance of limiting contamination and materials management problems, including the generation of wastes and contaminated media, in ongoing and future DOE operations.

DOE asked the committee to consider whether the R&D portfolio should address environmental problems outside of DOE that are related

---

[4] Throughout this report the committee uses the term "short-term" to mean 5 years or less, and "long-term" to mean greater than 5 years.

[5] The term "EQ problems" refers to the set of technical problems that collectively make up the scientific and technical challenges described earlier. This is a useful concept in planning an R&D portfolio because the challenges are very broad, and must be broken down into manageable parts to be addressed by R&D.

to EQ strategic goals. **The committee concludes that it is appropriate for the EQ R&D portfolio to address environmental problems outside DOE if such R&D is directly related to DOE's EQ mission. At this time, however, the EQ R&D portfolio should not address environmental problems beyond DOE's jurisdiction that are unrelated to the EQ mission.** There may be cases in which spending limited R&D resources on problems outside DOE's EQ mission is appropriate, but deciding when this would be appropriate is less a technical question than a matter of general policy.

## ADDRESSING LONG-TERM, CURRENTLY INTRACTABLE[6] EQ PROBLEMS

Many of the problems confronting the EQ business line are long-term, both because they involve materials that remain hazardous, in some cases, for thousands to hundreds of thousands of years and because they are so complex and unique that R&D may have to continue for decades to generate their solutions.[7] DOE is responsible for managing, removing (or isolating), and disposing of uniquely hazardous, chemically complex substances, such as spent nuclear fuel, liquid high-level radioactive wastes, nuclear materials, mixtures of hazardous and radioactive compounds, and a wide range of contaminated media (e.g., groundwater, soil, and nuclear production facilities). These activities must be carried out under a wide range of challenging and often unique circumstances. Environmental cleanup, waste management, and disposal activities will, of necessity, endure for generations and long-term stewardship at most DOE sites may continue indefinitely. The future should provide opportunities for continual improvements and possible breakthrough technologies that could greatly reduce risks to human health and the environment and costs to future generations. **The committee concludes that the uniqueness and complexity of DOE's EQ problems demand that the EQ R&D portfolio have a strong, if not dominant, long-term component. The committee recommends that DOE begin to devote an increasing fraction of its EQ R&D to long-term problems to ensure that an R&D portfolio dedicated to long term problems is in place within five years. The committee also recommends that DOE develop a strategic vision for its EQ R&D portfolio. This vision should provide the framework for developing the science and technology necessary to address EQ problems that**

---

[6] The committee uses the term "currently intractable" to refer to problems for which there are no identified, acceptable solutions but for which long-term R&D could lead to such solutions.
[7] When the expression "long-term R&D" is used in this report, the committee means "long-term" from both of these perspectives.

extend beyond the present emphasis of short-term "compliance" and should incorporate the principle of continual improvement.

### ADVANCING MORE INFORMED DECISION MAKING

Numerous decisions on environmental remediation, waste management, materials storage, and facility decommissioning involve complex technical issues for which there are only limited data and partial scientific understanding. These gaps in knowledge affect DOE's decisions on each of the EQ challenges listed above. It should be emphasized, however, that lack of technical information does not necessarily preclude effective decision making. There are a number of examples of long-term EQ challenges (e.g., long-term stewardship, and geological disposal) where current decisions should include consideration that technology and understanding can be expected to improve considerably during the timeframe of the challenge.

For residual contamination at closed legacy sites, for example, the system of long-term stewardship should not preclude future actions to address remaining risks to human health and the environment. The system should allow future decision makers to re-initiate active cleanup activities if and when future technologies improve to a point where it makes sense to address remaining risks, or when the understanding of the effects of DOE wastes and contaminated media on human health and the environment improve. For geological disposal of high-level wastes and spent nuclear fuel, DOE should pursue a phased approach that would allow changes to the disposal plans to improve operations, safety, schedule, or cost throughout the decades-long process of emplacement. Such a phased decision process also could be applied to other important long-term EQ problems. DOE's EQ R&D portfolio should support decision making by including R&D on technical alternatives in cases where existing techniques are expensive, inefficient, or pose high risks to human health or the environment, or where techniques under development have high technical risks.[8] **The committee concludes that the EQ R&D portfolio is critical to improving decision making and should be designed to help inform important DOE decisions, including support for technical alternatives in areas of high cost or high risk.**

---

[8] Technical risk is defined as "the probability that the technique or method fails to accomplish the goals and performance requirements set by policy or regulation."

## CRITERIA TO EVALUATE THE ADEQUACY OF THE EQ R&D PORTFOLIO

The committee presents its analysis of the important functions of an effective strategic EQ R&D portfolio in Chapter 3. These functions, and the accompanying findings, conclusions, and recommendations, were used to develop a set of criteria to evaluate the adequacy of the portfolio. **The committee recommends that DOE use, at a minimum, the following 10 criteria for this purpose:**

1. **There should be no significant gaps in critical areas of science and technology that are required to address EQ goals and objectives.**
2. **The portfolio should support the accomplishment of closely related DOE and national missions.**
3. **The portfolio should include R&D to develop technical alternatives in cases where: (1) existing techniques are expensive, inefficient, or pose high risks to human health or the environment; or (2) techniques under development have high technical risk.**
4. **The portfolio should produce results that could transform the understanding, need, and ability to address currently intractable problems and which could lead to breakthrough technologies.**
5. **The portfolio should leverage R&D conducted by other DOE business lines, the private sector, state and federal agencies, and other nations to address EQ goals and objectives.**
6. **The portfolio should help narrow and bridge the gap between R&D and application in the field.**
7. **The portfolio should be successful in improving performance, reducing risks to human health and the environment, decreasing cost, and advancing schedules.**
8. **There should be an appropriate balance between addressing long-term and short-term issues.**
9. **A diversity of participants from academia, national laboratories, other federal agencies, and the private sector, including students, postdoctoral associates, and other early-career researchers, should be involved in the R&D.**
10. **There should be an appropriate balance of annual new starts, extensions of promising R&D, and periodic new initiatives.**

## PRINCIPAL ELEMENTS OF THE EQ R&D PORTFOLIO

The committee was asked to advise DOE on the principal elements of its EQ R&D portfolio. The committee approached this task in two ways: (1) by developing its own list of elements and (2) by developing a general methodology that DOE could use to build upon this list of program priorities to achieve and maintain a more strategic EQ R&D portfolio. The committee's list of principal elements is presented below (and discussed more fully in Chapter 3), whereas a summary of the proposed methodology is presented in the next section (and discussed more fully in Chapter 4). In developing its list, the committee attempted to take a high-level, long-term view of the R&D needed to address DOE's most challenging EQ problems. Accordingly, the elements generally were not defined along existing DOE program lines and are quite broad.

**The committee recommends that DOE's EQ R&D portfolio include, at a minimum, the following 5 principal elements:**

**1. Development and evaluation of approaches that reduce the impacts of wastes on human health and the environment through generation minimization; processing improvements, including volume reduction, stabilization, and containment; and disposal.**
**2. Development of methods and techniques for cutting-edge characterization and remediation of contaminated media, including facilities.**
**3. Improvement of understanding of the movement and behavior of contaminants through the environment, with an emphasis on locating and tracking the movement of contaminants in the subsurface.**
**4. Development of mechanisms for effective long-term stewardship, including improved institutional management capabilities, appropriate monitoring, and the means to implement future improvements in technology and understanding.**
**5. Determination of the risks of DOE wastes and contaminated media to human health and the environment to improve the bases upon which regulatory and societal decisions can be made.**

## PORTFOLIO MANAGEMENT PROCESS

The committee has described a vision for an EQ R&D portfolio that emphasizes more strongly DOE's long-term EQ problems. To move towards this vision, DOE must redesign and rebalance its EQ R&D portfolio in substantial ways. In Chapter 4 the committee describes a portfolio management process that could help achieve these goals. For the most part DOE can implement the recommended new portfolio management process through an evolutionary approach, i.e., by modifying and supplementing existing management processes. The committee believes this is possible because DOE is already using portfolio management techniques and external reviews have found that management processes based on these techniques are yielding positive results but could be greatly improved. Such an approach avoids disruptive reorganizations and maintains management focus on the goal, i.e., realizing the new R&D vision. The key elements of this process are discussed briefly below.

## Generating an Improved Set of R&D Project Ideas

DOE's present R&D planning processes for the EQ portfolio are designed primarily to gather current information needs of the sites (i.e., EM cleanup sites or repositories), which tend to be focused primarily on short-term problems and the R&D to address them. Most of the participants in these processes are DOE employees and contractors who are involved in the site problems and issues, with some periodic input from the broader technical community. The existing R&D planning processes are unlikely to generate the full scope of strategic R&D needed to address DOE's most challenging, long-term EQ problems. **The committee recommends that DOE establish a new mechanism within its portfolio management process whose purpose is to develop a more strategic EQ R&D portfolio. This new process, termed the "Strategic Portfolio Review," should supplement and operate in parallel with existing site-driven processes. The Strategic Portfolio Review should be carried out by an independent planning and review board specifically focused on the EQ R&D portfolio, with membership composed of leaders in the scientific and technical community, including experts from industry, academia, national laboratories, and affected communities.** The expanded set of EQ R&D projects to be considered for funding would consist of projects emerging from the traditional needs processes as well as from the new Strategic Portfolio Review.

## Measuring the Magnitude of the Benefit

The committee found that DOE does not have a method for prioritizing R&D activities across the entire EQ portfolio. Each DOE organization that supports EQ R&D activities has its own process for prioritizing and selecting R&D activities. The process used to select over 80 percent of the EQ R&D portfolio (those activities supported by EM) is EM's Work Package Ranking System. The current Work Package Ranking System is strongly biased toward activities that are site-generated and connected to the present remediation plans. Moreover, it is, by design, EM specific and therefore does not apply to other parts of the EQ R&D portfolio. The current ranking system is unlikely to be effective in prioritizing R&D activities designed to address the strategic gaps and opportunities identified in the Strategic Portfolio Review discussed above, especially those not within EM. **The committee recommends that DOE develop and implement an evaluation method to address more strategic R&D for the entire EQ R&D portfolio. In the short term, it could be entirely separate from EM's Work Package Ranking System, but, in the longer term, a new approach is needed that works for both site-driven activities and strategy-driven activities and is applied within all areas (i.e., EM, RW, NE) of the EQ R&D portfolio.** Several useful non-EM-specific models that have been applied to elements of the EQ R&D portfolio are discussed in Chapter 4.

## R&D Centers

After identifying important strategic R&D activities through the processes described above, it is essential for DOE to provide longer-term support for them, specifically countering the "going out of business within the next decade" philosophy that has permeated some views of the EQ mission. The committee believes that a significant fraction of R&D should be conducted in organizationally separate units to help maintain a focus on long-term results. Each of these units would be strongly coupled to an important, currently intractable EQ problem and evaluated according to progress on solving the problem, but not strongly coupled to short-term program needs. **The committee recommends that DOE implement a new approach to provide longer-term funding for organizationally separate, integrated, and coordinated R&D activities (i.e., R&D centers[9]) designed to solve well-defined, high-priority EQ problems.** Chapter 4 provides details on how DOE could implement this approach.

## DETERMINING AN APPROPRIATE LEVEL OF R&D INVESTMENT

EQ is DOE's most expensive business line, accounting for approximately $6.7 billion, or 36 percent of DOE's total budget. In contrast, the annual investment in EQ R&D is the smallest of DOE's programmatic business lines, accounting for only 4 percent of DOE's total R&D spending. These budget data are an indication that decision makers in DOE, the Office of Management and Budget, and Congress may not fully understand the magnitude and duration of many of the challenges faced by the EQ business line, and the potential value of long-term R&D to ad-

---

[9] The committee refers to the organizations carrying out the integrated and coordinated R&D efforts as "R&D centers" to indicate that the whole of each is greater than the sum of its parts. This synergy could be achieved in more than one way (see discussion in Chapter 4).

dress such challenges.

The appropriate level of R&D funding depends on the scope of the EQ mission and must take into account the balance between spending limited resources on R&D and other possible uses of those resources. Broad-based support for R&D requires a compelling commitment to the goals and objectives of DOE's EQ mission. The committee recommends that DOE develop new strategic goals and objectives for its EQ business line that explicitly incorporate a more comprehensive, long-term view of its EQ responsibilities. After clear goals and objectives have been defined, DOE managers and others will have to deal with difficult tradeoffs in determining the level of R&D funding. There are many important short-term problems that call for high-priority allocation of funds. Often reinforcing or driving these needs are milestones associated with existing compliance agreements between DOE and state environmental regulatory authorities, congressional expectations, and concomitant expectations of the affected communities and their representatives. In such situations, allocating funds to R&D can be seen as taking resources from meeting short-term requirements or compliance agreements to support activities that are, by their very nature, longer term and more uncertain in their ultimate benefits. It is, therefore, incumbent upon DOE leadership to make clear to all EQ stakeholders the value of a strong and sustained R&D portfolio in addressing long-term EQ problems.

It has not been possible to identify an analytic or quantitative approach to establish an appropriate level of EQ R&D funding. There are two general techniques that, together, could be used for this purpose: (1) benchmarking against other mission-driven R&D efforts, both nationally and internationally, and (2) applying a set of investment indicators based closely on the adequacy criteria developed earlier.

Benchmarking the level of EQ R&D funding against similar programs could provide a meaningful measure for discerning a range of reasonable R&D investment levels. It also could help to explain and justify the level of future EQ budget requests to decision makers within DOE, the Office of Management and Budget, and Congress and to other interested parties. **The committee recommends that DOE benchmark the EQ R&D budget against other mission-driven federal R&D programs in the federal government. Such benchmarking exercises should have participation or review by outside experts. Proposed budgets should be presented in the context of benchmarking, and significant deviations from the information gained through benchmarking should be explained.**

The 10 criteria described earlier to evaluate the adequacy of the EQ R&D portfolio also can be used as guides for determining an appropriate level of investment. The committee's list of these 10 investment indicators is provided in Chapter 5. Meeting such criteria is an important indication of an appropriately formulated R&D portfolio. Although the level of

R&D investment alone cannot guarantee the achievement of these indicators, the level of investment should not preclude their achievement. **The committee recommends that DOE use investment indicators, together with benchmarking techniques, to help determine the appropriate level of EQ R&D investments.**

## CONCLUSION

DOE's EQ R&D portfolio must be recognized as centrally important to DOE's EQ and other missions, and an enduring responsibility of the department. R&D success requires an adequate, stable, and predictable level of funding. A well-designed, sufficiently funded, and well-implemented EQ R&D portfolio is necessary, but not sufficient, to assure that the potential value of R&D in addressing DOE's EQ problems is achieved. Many other features must be present, including technically competent and trusted R&D program managers; effective relationships among problem holders, R&D managers and researchers; good communication of R&D results; and incentives for R&D results to be used in solving problems.

An effective portfolio also requires close and trusting relationships among the responsible DOE headquarters and local officials, contractors at the sites, state regulatory officials, and stakeholders such as the affected community. The nature of successful EQ R&D is to present opportunities to reduce risks to workers and the public, improve schedules, decrease costs, and solve problems. But it also can require re-addressing existing agreements, changing schedules, dealing with periods of uncertainty, and revisiting expectations. All of these factors must be resolved for DOE's EQ R&D to achieve its goals. An EQ R&D portfolio that is well conceived, effectively managed, adequately and consistently funded, and championed by DOE leadership is essential to success in achieving the DOE EQ mission.

# 1

# INTRODUCTION

The U.S. Department of Energy (DOE) is responsible for a diverse range of radioactive, hazardous, and mixed low-level wastes; nuclear materials; spent nuclear fuels; and contaminated lands, waters, and facilities (hereafter referred to collectively as "DOE wastes and contaminated media," [see Sidebar 1.1]). These wastes and contaminated media present the following general scientific, technical, and social challenges that will endure long into the future (see Chapter 2 for a more complete discussion of these challenges):

- Remediate DOE sites[1] and facilities that have severe radioactive and hazardous waste contamination from past activities (also commonly referred to as site "cleanup," see Sidebar 1.2).
- Manage, stabilize, process, and dispose of a legacy of widely varying and often poorly understood DOE wastes (including spent nuclear fuels and nuclear materials treated as waste) that are potential threats to health, safety, and the environment.
- Provide effective long-term stewardship of DOE sites that have been remediated as well as currently practical, but that have residual risks to health, safety, and the environment (see Sidebar 1.2 for definition of "long-term stewardship").
- Develop, open, and operate unique, first-of-a-kind facilities for the permanent disposal of radioactive spent fuels and high-level wastes—many of which will be hazardous for thousands to hundreds of thousands of years.
- Limit contamination and materials management problems, including the generation of wastes and contaminated media, in ongoing and future DOE operations.

---

[1] See Sidebar 2.2 for a description of the largest DOE sites.

**SIDEBAR 1.1 DOE Wastes and Contaminated Media**

Radioactive wastes are the unwanted byproducts of the nuclear fuel cycle and can contain both radioactive isotopes and hazardous chemicals. In the United States, radioactive waste is classified and managed by its source of production rather than by its physical, chemical, or radioactive properties. Consequently, different classes of waste can contain many of the same radioactive isotopes, and even low-level waste can contain certain long-lived radioactive isotopes.

In general, nuclear fuel cycle wastes are grouped into the following broad classes for purposes of management and disposal:

• *High-level waste* is the primary waste produced from chemical processing of spent nuclear fuel. This waste is usually liquid and contains a wide range of radioactive and chemical constituents, and must be solidified before permanent disposal.
• *Spent nuclear fuel* is fuel that has been irradiated in a nuclear reactor, and for the purposes of disposal can include cladding and other structural components.
• *Transuranic waste* excludes high-level waste as defined above and includes waste that contains alpha-emitting transuranium (i.e., atomic number greater than 92) isotopes with half-lives greater than 20 years and concentrations greater than 100 nanocuries per gram. This waste usually consists of contaminated materials like clothing and tools used in the manufacture of nuclear weapons.
• *Mill tailings* are wastes resulting from the processing of ore to extract uranium and thorium.
• *Low-level waste* is radioactive waste that does not meet one of the definitions given above.
• *Mixed low-level waste* is low-level waste that contains both chemically hazardous and radioactive components.

There are two other classes of materials that DOE sometimes manages as waste:

• *Nuclear materials*, such as plutonium and special-use isotopes, that may be declared surplus and disposed of as waste.
• *Contaminated media,* such as contaminated soil, groundwater, and buildings, whose cleanup may generate additional radioactive and chemical waste streams that must be treated and managed.

The term "DOE wastes" is used throughout this report to refer to all wastes and spent nuclear fuels and nuclear materials treated as waste described above for which DOE is responsible. Similarly, the term "DOE wastes and contaminated media" is used to encompass all wastes types, spent nuclear fuels, nuclear materials, and contaminated media described above for which DOE is responsible.

---

**SIDEBAR 1.2 Definitions of *Cleanup* and *Long-Term Stewardship***

**Cleanup**

DOE defines *cleanup* as the "process of addressing contaminated land, facilities, and materials in accordance with applicable regulations. Cleanup does not imply that all hazards will be removed from the site" (DOE, 1999a).

According to DOE, site cleanup is complete when the following five criteria have been met:

1. Deactivation or decommissioning of all facilities currently in the EM program have been completed, excluding any long-term surveillance and monitoring.
2. All releases to the environment have been cleaned up in accordance with agreed-upon cleanup standards.
3. Groundwater contamination has been contained and long-term treatment (remedy) or monitoring is in place.
4. Nuclear materials have been stabilized and/or placed in safe long-term storage.
5. Legacy waste has been disposed of in an approved manner (legacy waste was produced by past nuclear weapons production activities. (DOE, 1998).

As DOE has stated, "completing cleanup," also commonly referred to as site "closure," does not mean that the site will be made available for unrestricted use. In fact, most DOE sites will require some form of long-term stewardship (see below) to protect human health and the environment from hazards after cleanup is complete.

**Long-Term Stewardship**

DOE defines long-term stewardship as "all activities necessary to ensure protection of human health and the environment following completion of cleanup, disposal, or stabilization at a site or a portion of a site. Long-term stewardship includes all engineered and institutional controls designed to contain or to prevent exposures to residual contamination and waste, such as surveillance activities, record-keeping activities, inspections, groundwater monitoring, ongoing pump and treat activities, cap repair, maintenance of entombed buildings or facilities, maintenance of other barriers and containment structures, access control, and posting signs" (DOE, 2001c).

---

DOE currently spends approximately $6.7 billion a year on activities designed to address these challenges. DOE refers to these activities as its Environmental Quality (EQ) "business line."

The magnitude and duration of these challenges are related to:

- the quantities of DOE wastes and contaminated media that are distributed at numerous sites throughout the United States;
- the quantities of wastes and contaminated media currently outside DOE (or to be generated in the future) that are expected to be added to DOE's current inventories;
- the long half-lives (thousands to hundreds of thousands of years) of some of the radioactive elements contained in these materials; and
- the potential risks to humans and the environment if radioactive materials (and related hazardous substances) are not adequately isolated from the biosphere.

The long-term nature of these challenges and the enormous costs associated with them create many opportunities to improve methods, lower costs, reduce impacts to human health and the environment, and improve stewardship through a strengthened long-term research and development (R&D) effort (see Sidebar 1.3 for definitions of "research" and "development"), as discussed in more detail in Chapter 3. One direct effect could be to reduce the costs of managing and disposing of DOE wastes and contaminated media through the development of more cost-effective approaches. Long-term R&D also could lead to novel approaches to reduce the risks of DOE wastes and contaminated media to human health and the environment to levels that are not possible given current technical capabilities and understanding. Another significant potential impact of long-term EQ R&D is improvement in technical understanding of issues related

---

**SIDEBAR 1.3 Definitions of "Research" and "Development"**

The committee has adopted the definitions of "research" and "development" that are used by the National Science Foundation, the Office of Management and Budget, and federal agencies (including DOE) to report R&D funding data. The term "research" is defined as systematic study directed toward more complete scientific knowledge or understanding of the subject studied. The term "development" is defined as the systematic use of the knowledge or understanding gained from research for the production of materials, devices, systems, or methods, including design, development, and improvement of prototypes and new processes. It excludes quality control, routine product testing, and production. Therefore, throughout this report the short-hand "R&D" is used to include all stages of technology maturation from research through demonstration and initial deployment of a new technology, rather than the more precise term "research, development, demonstration, and deployment", or "RDD&D."
Source of Definitions: American Association for the Advancement of Science, http://www.aaas.org/spp/dspp/re/define.htm.

to cleanup, long-term stewardship, and the disposal of radioactive wastes, which could help policy makers make more informed decisions.

R&D activities designed to address these challenges are supported principally by three offices: the Office of Environmental Management (EM), the Office of Civilian Radioactive Waste Management (RW), and the Office of Nuclear Energy, Science and Technology. In addition, the Office of Fissile Materials Disposition conducts a small amount of R&D as part of the EQ business line, and DOE's Office of Science, which is not formally part of the EQ business line, supports a variety of basic research activities related to DOE's EQ mission. A brief overview of the types of R&D supported by these offices is provided in Sidebar 1.4.

## DOE'S R&D PORTFOLIO PROCESS

In 1998 DOE began an effort under the direction of the Under Secretary to develop comprehensive descriptions of its R&D activities. These descriptions have been organized into R&D portfolio documents that describe the collection of R&D activities (i.e., the R&D portfolio[2]) which supports each of the four programmatic business lines[3] established in DOE's strategic plan (DOE, 1997b, 2000f): EQ (DOE, 2000b), Energy Resources (DOE, 2000a), National Nuclear Security (DOE, 2000c), and Science (DOE, 2000d). The primary objective of this portfolio-development effort was to ensure that R&D activities are focused on the goals outlined in DOE's strategic plan. DOE stated that it would use these portfolios to better manage its R&D programs, most notably in the following ways:

- to increase coordination and integration of R&D activities across the department;
- to identify R&D gaps and opportunities;
- to balance R&D investments; and
- to provide a coherent rationale for R&D budget requests to the Office of Management and Budget and the U.S. Congress.

Although DOE's motivation for the portfolio reviews did not include a need to examine DOE's R&D responsibilities in the context of other U.S. and international agencies, the committee believes that this is an important issue to consider as well. The role of the EQ R&D portfolio in ad-

---

[2] The committee uses the term "portfolio" to refer to the collection of R&D activities that support each of DOE's four programmatic business lines, and the term "portfolio document" to refer to the published description of these activities.

[3] The four programmatic business lines are supported by a corporate management function, which DOE's most recent strategic plan refers to as a fifth business line (DOE, 2000f).

dressing non-DOE problems, both national and international, is an explicit charge to this committee (see below).

## STATEMENT OF TASK

The Under Secretary of Energy asked the National Academies' National Research Council to address the following questions related to the EQ R&D portfolio, with a focus on post-2006 research:

- In the context of EQ strategic goals and mission objectives, what criteria should be used to evaluate the adequacy of the portfolio?
- Using these criteria, what should be the principal elements of the portfolio?
- Should the portfolio be designed to address environmental problems outside DOE (e.g., Department of Defense, Russia) that are related to EQ strategic goals?
- How to determine the level of future investments in EQ R&D?

One issue that the committee had to address in carrying out the study was how to interpret the phrase "with a focus on post-2006 research." In 1996 EM established a formal goal of completing cleanup at as many sites as possible by the year 2006 (DOE, 1998). Because this goal focuses exclusively on the number of sites "cleaned up" by 2006 (see Sidebar 1.2), it has had a major effect on EM's approach to remediation activities. In particular, the goal created an incentive to focus resources on sites that could realistically be cleaned up within a 10-year or shorter time frame relative to larger sites faced with more difficult problems. It also may have given the incorrect perception to some federal decision makers that environmental problems associated with DOE wastes and contaminated media would largely be addressed by 2006 (see discussion in Chapter 2). The short-term focus of the goal has had a marked impact on the types of R&D activities supported by EM: It created a clear incentive to support late-stage development and deployment of techniques that could be used in the near term to facilitate cleanup of sites by 2006, and a disincentive to fund long-term R&D activities that might address the more difficult problems that will endure beyond 2006. These factors, together with other external forces driving EM to produce results quickly (i.e., congressional expectations, regulatory constraints), have resulted in a portfolio of EM R&D activities heavily weighted toward short-term needs. It also may partly account for the recent trend of decreasing investments in EQ R&D (see discussion in Chapter 5).

Although the year 2006 has no special meaning for the other DOE organizational units involved in the EQ business line, RW also has been driven to focus its R&D activities on short-term needs—most notably, sci-

## SIDEBAR 1.4 Department of Energy Offices That Support EQ R&D

Five DOE offices support R&D related to DOE's EQ business line. The missions of these offices, their significant EQ R&D efforts, and their total EQ R&D funding levels are summarized briefly below.

### Office of Environmental Management (EM)

In 1989, Congress established EM to reduce threats to health and safety posed by environmental contamination at DOE sites. Most of the R&D activities considered part of the EQ R&D portfolio are supported by EM's Office of Science and Technology (OST) and by the Environmental Management Science Program (EMSP), which is administered jointly by EM and the Office of Science. Activities conducted by the sites themselves to develop new techniques or refine existing techniques are not considered part of DOE's EQ R&D portfolio because they are integral parts of large operating projects (see Sidebar 2.2 for more information on DOE sites).

OST's mission is to manage and direct a national, solution-oriented science and technology program to provide the scientific foundation, new approaches, and new technologies that could significantly reduce the risk, cost, and time needed for completion of the EM cleanup mission. To accomplish this mission, OST supports technology-development activities in five focus areas (deactivation and decommissioning, nuclear materials, subsurface contamination, tanks, and transuranic and mixed low-level waste) and in five crosscutting areas (characterization, monitoring, and sensor technology; efficient separations; industry/university programs; long-term stewardship; and robotics). EMSP supports mission-driven basic research of relevance to EM's cleanup mission. EM's EQ R&D budget in fiscal year (FY) 2001 is approximately $240 million.

### Office of Civilian Radioactive Waste Management (RW)

The Nuclear Waste Control Act of 1982 established RW to develop and manage a federal system for disposing of all spent nuclear fuel from commercial nuclear reactors and high-level radioactive waste and spent nuclear fuel resulting from atomic energy defense activities. The Nuclear Waste Policy Amendments Act of 1987 directed DOE to characterize only Yucca Mountain, Nevada, to determine its suitability as a repository site for the disposal of spent nuclear fuel and high-level radioactive waste. RW's R&D activities address the following issues: understanding the effects of heat on repository system performance, building a three-dimensional model of the Yucca Mountain site, enhancing repository design, improving the design of waste packages and drip shields, developing dry transfer systems for spent nuclear fuel, conducting performance assessments for various repository conditions, and improving the understanding of the saturated zone beneath Yucca Mountain. RW's EQ R&D budget in FY 2001 is approximately $45 million.

*continues on next page*

**Office of Nuclear Energy, Science, and Technology (NE)**

NE is responsible for managing the federal government's investment in nuclear science and technology and supporting innovative applications of nuclear technology. Most of NE's R&D activities considered part of the EQ R&D portfolio address the management of existing inventories of depleted uranium hexafluoride[a] and sodium-bonded spent nuclear fuel. NE sponsors investigator-initiated, peer-reviewed research at universities, national laboratories, and industry through its Nuclear Energy Research Initiative (NERI). One component of NERI is to support research on advanced spent fuel treatment technologies (such as electrometallurgical treatment technologies) that could reduce the volume of spent nuclear fuel and other radioactive waste. At the direction of Congress, NE also is examining the feasibility of accelerator transmutation of waste technologies, which offer the potential to reduce the amount of long-lived radionuclides in waste by transforming plutonium, long-lived actinides, and long-lived fission products contained in spent fuel. NE's EQ R&D budget in FY 2001 is approximately $11 million.

**Office of Fissile Materials Disposition (MD)**

MD is responsible for all activities of DOE's National Nuclear Security Administration relating to the management, storage, and disposition of fissile materials from weapons and weapon systems that are excess to U.S. security needs. MD EQ R&D focuses on developing techniques to transform weapons-usable plutonium to forms that are not readily accessible for use in nuclear weapons. MD's EQ R&D budget in FY 2001 is approximately $3 million.

**Office of Science (SC)**

SC funds basic research organized around four main themes: (1) fueling the future (including research on new fuels, and clean and affordable power), (2) protecting our living planet (including energy impacts on people and the environment); (3) exploring matter and energy, and (4) extraordinary tools for extraordinary science (including national assets for multidisciplinary research). Although SC is not formally part of DOE's EQ R&D portfolio, it supports a variety of research activities related to DOE's EQ mission, primarily within its Office of Basic Energy Sciences (BES) and the Office of Biological and Environmental Research (BER). BES research efforts that could have an impact on DOE's EQ mission include some elements of the geoscience, separations science, and materials science and engineering programs. BER research efforts that could have an impact on DOE's EQ mission include some elements of its natural and accelerated bioremediation program and its Environmental Molecular Sciences Laboratory.

---

[a] In October 2000 the depleted uranium cleanup program was transferred from NE to EM.

entific investigations to assess the suitability of the proposed Yucca Mountain repository, which if approved and licensed, could begin receiving waste as early as 2010—even though the program will endure and should benefit from R&D for generations if successful. Therefore, by asking the committee to focus on post-2006 R&D activities, DOE was looking for advice on how the EQ R&D portfolio could shift its current short-term focus to a more long-term, strategic view. Throughout this report the committee uses the term "short-term" to mean 5 years or less, and "long-term" to mean greater than 5 years.

Finally, part of the committee's task is to provide "criteria ... to evaluate the adequacy of the portfolio" and to "identify the principal elements of the portfolio." To this end Chapter 3 describes a long-term vision for the EQ portfolio, and Chapter 4 describes how DOE could achieve the vision. One might say that, in the narrowest sense of the term "criteria," Chapter 3 fulfills the task and Chapter 4 exceeds it. However, the committee believes that it is important to describe more than where the portfolio should be when it is "adequate," for doing only that would provide little practical guidance today on steps to move in the correct direction. Describing the characteristics of an adequate portfolio is necessary but not sufficient. It is also necessary to describe a process that will achieve and maintain an adequate portfolio. This is the proverbial difference between giving a man a fish and teaching him to fish. In sum, one criterion for an adequate portfolio is a process to develop and maintain it.

## STUDY PROCESS

The National Academies appointed a committee of twelve experts[4] with a range of perspectives and experience related to the task (see Appendix A for biographical information on committee members). The committee held its first information-gathering meeting on June 7 and 8, 2000, in Washington, D.C. During this meeting, it discussed its task with DOE leadership and heard presentations from a number of DOE program managers. The primary information-gathering activity for the study was a two-day public workshop held in Washington, D.C. on August 23 and 24, 2000. This workshop brought together approximately 50 participants from DOE, the private sector, academia, and other federal agencies (see Appendix B for a list of participants and the workshop agenda). The workshop began with a series of keynote speakers who discussed various aspects of DOE's EQ R&D activities. Participants were then organized into three working groups to examine issues associated with: (1) identifying

---

[4] Teresa Fryberger, who served as vice-chair of the committee, recused herself from committee activities in November 2000 and resigned from the committee in January 2001 after accepting a management position within DOE-EM.

significant long-term EQ R&D needs; (2) evaluating the balance and value of the EQ R&D portfolio; and (3) determining the appropriate level of investment in long-term EQ R&D. Prior to the workshop, the committee also solicited input on significant long-term R&D needs from persons knowledgeable about DOE's EQ mission, and these issues were used as input to the workshop. The committee then held two closed meetings during which it developed its findings, conclusions, and recommendations and prepared this report.

This study complements and builds on two other recent analyses of DOE's EQ R&D portfolio. Last year DOE's Strategic Laboratory Council[5] conducted an adequacy analysis to examine the capability of the current portfolio of DOE R&D activities to meet the objectives of the EQ business line (DOE, 2000g). Based on an extensive review of the current portfolio, the adequacy analysis recommended a number of changes to DOE's EQ strategic goal and objectives, identified a large number of R&D gaps and opportunities,[6] and offered findings and recommendations on how the portfolio could be improved. After the adequacy analysis was published, the Technology Development and Transfer Committee of DOE's Environmental Management Advisory Board (EMAB)[7] was asked to evaluate the analysis, resulting in an EMAB letter report in October 2000 (DOE, 2000g). The results of the adequacy analysis and the EMAB letter report are summarized in Appendix C.

## ORGANIZATION OF REPORT

This chapter has provided an introduction and general background on the issues addressed in the report. Chapter 2 describes the mission of DOE's EQ business line, summarizes its major areas of responsibility, and

---

[5] The Strategic Laboratory Council (SLC) consists of representatives from DOE's national laboratories. It provides assistance to EM offices and EM-related science organizations on science and technology issues dealing with EM's program. Operating in a consensus manner, the SLC develops positions and offers recommendations that represent the DOE laboratory system. SLC members took the lead in organizing panels to conduct the adequacy analysis of the EQ R&D portfolio.

[6] According to the SLC's adequacy analysis, an R&D gap exists where "the current portfolio is less than adequate in some respect (e.g., lacking needed work), thus posing a risk of failure to achieve an EQ objective." An R&D opportunity exists "when there is significant potential to achieve a high return on investment or to excel in achieving an EQ objective through a new research area or more investment in an existing research area" (DOE, 2000g).

[7] EMAB is an advisory group chartered to provide advice to DOE's Assistant Secretary for EM on issues related to environmental restoration and waste management issues. EMAB members are chosen to represent key stakeholder groups in EM's decision-making process. EMAB carries out much of its work through its committees, including the Technology Development and Transfer Committee (which focuses on technology-related issues) and the Science Committee (which focuses on the quality of science in the EM program). Also see Sidebar 4.1.

discusses issues associated with the scope of the EQ mission. Chapters 3 and 4 together provide the committee's vision for a more effective long-term EQ R&D portfolio. Chapter 3 begins by discussing the important functions of such a portfolio, which are used as the basis of the committee's list of criteria to evaluate the adequacy of the portfolio that follows. Based on these criteria and the findings of many recent studies, the committee develops a short list of principal elements, presented at the end of Chapter 3, that it believes are going to be essential to the success of DOE's long-term EQ mission. Chapter 4 then describes how DOE could build upon the adequacy criteria and principal elements developed in Chapter 3 to achieve and maintain a more strategic, long-term R&D port-folio. Finally, Chapter 5 describes processes that could be used to help determine an appropriate level of investment in EQ R&D. Supporting materials are included as Appendixes A through F.

# 2

# THE DEPARTMENT OF ENERGY'S
# ENVIRONMENTAL QUALITY MISSION

The Department of Energy's (DOE's) Environmental Quality (EQ) business line encompasses some of the largest, costliest, and most complex environmental remediation, nuclear and hazardous materials management, and waste disposal programs in the world (OTA, 1991, 1993a,b; DOE, 1995). In this chapter the committee focuses on the mission of the business line, rather than its research and development (R&D) activities. The purpose of the chapter is to provide the context that is required for the analyses of the R&D portfolio in the rest of the report. To do so, the committee provides a high-level overview of the major areas of responsibility of the business line and important elements of its budget. The committee also discusses and clarifies its views on the topical and temporal breadth of DOE's EQ mission.

## DOE'S EQ RESPONSIBILITIES

A general sense of the magnitude and enduring nature of DOE's EQ responsibilities can be ascertained from the quantities of DOE wastes and contaminated media[1] involved, the estimated life-cycle costs[2] for protecting human health and the environment from these wastes and contaminated media, and estimates of how long it will take to address these issues using existing technologies and current technical understanding. Table 2.1 summarizes these data, as well as annual R&D budgets, for each of the ten technical categories of the EQ business line. (More detailed descriptions of these technical categories are provided in Appendix D.) Some of the most significant characteristics of DOE's envir-

---

[1] See Sidebar 1.1 for definition of "DOE wastes and contaminated media."

[2] All life-cycle cost data in this chapter are DOE's estimates of the costs of addressing these problems at all DOE sites from fiscal year 1997 through 2070 in 1999 dollars (DOE, 2000e). This total incudes some costs already incurred in fiscal years 1997 through 2000. The committee has not validated the accuracy of these estimates.

TABLE 2.1 Summary of Major Technical Categories Addressed by DOE's EQ R&D Portfolio

| Technical Category[a] | Quantities | Major Sites | Life-Cycle Costs[b] | Estimated Duration of Problem | Annual R&D Budget (FY 2001 request)[b] | Responsible DOE Office[f] |
|---|---|---|---|---|---|---|
| Manage high-level waste | 340,000 $m^3$ stored in 280 large tanks and 63 smaller tanks | Savannah River, Hanford, INEEL, Oak Ridge, West Valley | $53.5 billion | Decades | $62 million | EM |
| Manage mixed low-level waste (MLLW) and transuranic (TRU) waste | 165,000 $m^3$ at 36 sites; an additional 45,000 $m^3$ of TRU and 170,000 $m^3$ of MLLW will be generated in next 10 years | 36 sites | More than $18 billion | Decades | $37.8 million | EM |
| Manage spent nuclear fuel | 2,100 MTHM[c] - Hanford 330 MTHM - INEEL 50 MTHM - Savannah River 85,000 MTHM commercial spent fuel | Hanford, INEEL, Savannah River | $7.8 billion | Decades | $36.3 million | EM, NE |
| Manage nuclear material | 200 metric tons of weapons-usable fissile materials; >700,000 metric tons of uranium hexafluoride | Rocky Flats Others | $6.4 billion | Decades | $24.4 million | EM, NE, MD |

| | | | | | | |
|---|---|---|---|---|---|---|
| Dispose high-level waste, spent nuclear fuel, and nuclear materials | 85,000 MTHM of commercial spent fuel; 22,000 canisters of high-level waste; 2,500 MTHM of DOE spent fuel | Yucca Mountain | $52-57 billion | Decades | $44.6 million | RW |
| Dispose TRU, low-level, mixed low-level, and hazardous wastes | 165,000 m$^3$ of mixed low-level and TRU waste at 36 sites; an additional 45,000 m$^3$ of TRU and 170,000 m$^3$ of MLLW will be generated in next 10 years | WIPP (TRU); Hanford, Savannah River, INEEL, Envirocare (low-level waste) | $8.1 billion for WIPP | Estimated closure date for WIPP is 2039 | No separate R&D budget | EM |
| Environmental remediation | 3 million m$^3$ of buried, solid radioactive and hazardous waste; 75 million m$^3$ of contaminated soil; 1.8 billion m$^3$ of contaminated groundwater | INEEL, Oak Ridge, Hanford, Rocky Flats, Savannah River | $12.6 billion | Cleanup estimated to continue through 2070 | $58.8 million | EM |
| Deactivation and decommissioning | 2,700 buildings containing over 180,000 MTHM | All | More than $37 billion | Cleanup estimated to continue through 2070 | $21.7 million | EM |

*continues on next page*

TABLE 2.1 Continued

| Technical Category[a] | Quantities | Major Sites | Life-Cycle Costs[b] | Estimated Duration of Problem | Annual R&D Budget (FY 2001 request)[b] | Responsible DOE Office[f] |
|---|---|---|---|---|---|---|
| Long-term stewardship | 129 DOE sites will require some form of long-term stewardship | 129 sites | Currently spends $64 million/year. By 2050, costs are estimated to be nearly $100 million/year[d] | Indefinite, potentially thousands of years | No separate R&D budget | EM |
| Minimize waste Generation | Not applicable | All | Estimates not available | Continuing | No separate R&D budget | All |

[a] Technical categories are those from the Strategic Laboratory Council's Adequacy Analysis (DOE, 2000g). See Appendix D for descriptions of technical categories.
[b] Data from Strategic Laboratory Council's Adequacy Analysis (DOE, 2000g).
[c] metric tons heavy metal.
[d] Estimates from DOE (2001b).
[e] Waste Isolation Pilot Plant.
[f] Office of Environmental Management (EM), Office of Civilian Radioactive Waste Management (RW), Office of Nuclear Energy, Science and Technology (NE); Office of Fissile Materials Disposition (MD).

onmental cleanup, materials management, and waste disposal responsibilities are discussed below.

## Addressing Environmental Contamination at DOE Sites

The DOE Office of Environmental Management (EM) is responsible for addressing environmental contamination problems at 140 sites located in 31 states (DOE, 1998; 1999a), which are referred to collectively as "the DOE complex" (see Sidebar 2.1). To date, DOE has identified almost 10,000 individual locations at these sites where toxic or radioactive substances were improperly abandoned or released directly into soil, groundwater, or surface waters (DOE, 1997a). An estimated 75 million cubic meters (2.6 billion cubic feet) of contaminated soil and 1.8 billion cubic meters (475 billion gallons) of contaminated groundwater may need to be remediated at these sites (DOE, 1997a). EM also is responsible for the deactivation and decommissioning of 2,700 facilities (of a total of about 20,000) determined to be surplus in the DOE complex (DOE, 1997a). Most of these facilities are seriously contaminated with radioactive or hazardous substances at levels that prohibit unrestricted release. Environmental remediation and deactivation and decommissioning activities are expected to continue at some sites through 2070, at a total life-cycle cost of nearly $50 billion for the entire DOE complex (DOE, 2000g).

## Managing DOE Wastes, Spent Nuclear Fuels, and Nuclear Materials

EM is responsible for more than 36 million cubic meters (9.6 billion gallons) of hazardous or radioactive wastes (DOE, 1997a), including over 340,000 cubic meters (90 million gallons) of high-level radioactive waste. Managing these wastes is extremely challenging, because of the volumes at issue, their hazardous characteristics, their long periods of toxicity and because much of it (including some of the most dangerous) is in unstable configurations (e.g., the Hanford tanks), much has been released to the environment already, and much of the waste is at present very incompletely characterized.

EM and DOE's Office of Nuclear Energy, Science and Technology (NE) also are responsible for managing over 800 million kilograms (1.8 billion pounds) of non-waste materials in inventory, such as depleted uranium, plutonium spent nuclear fuel, lead, sodium, lithium, and a variety of chemicals (DOE, 1996). Most of these materials, the majority of which is depleted uranium, are stored at 44 facilities at 11 major production sites throughout the United States. DOE's radioactive waste (including spent nuclear fuel and nuclear materials treated as waste) man-

**SIDEBAR 2.1 The DOE Complex**

Although the DOE complex encompasses over 100 distinct sites, the largest volumes of DOE wastes and contaminated media and many of the most costly and most challenging EQ problems are found at the six largest sites described below. The estimated site closure date and life-cycle costs for each site are those from EM's recent report, *Status Report on Paths to Closure* (DOE, 2000e).

1. The **Hanford Site** is in southeastern Washington State and covers an area of about 1,450 square kilometers (560 square miles). Production of materials for nuclear weapons took place here from the 1940s until mid-1989. The site contains several shutdown production reactors, chemical separations plants, and solid and liquid waste storage sites. **Estimated closure date: 2046. Estimated life-cycle costs: $56 billion.**

2. The **Savannah River Site**, near Aiken, South Carolina, covers an area of about 800 square kilometers (300 square miles). The site was established in 1950 to produce special radioactive isotopes (e.g., plutonium-239 and tritium) for use in the production of nuclear weapons. The site contains shutdown production reactors, chemical processing plants, and solid and liquid waste storage sites. **Estimated closure date: 2038. Estimated life-cycle costs: $37 billion.**

3. The **Idaho National Engineering and Environmental Laboratory**, first established as the Nuclear Reactor Testing Station and then the Idaho National Engineering Laboratory, occupies 2,300 square kilometers (890 square miles) in a remote desert area along the western edge of the upper Snake River plain. The site was established as a building, testing, and operating station for various types of nuclear reactors and propulsion systems, and the site also manages spent fuel from the naval reactor program. **Estimated closure date: 2050. Estimated life-cycle costs: $21 billion.**

4. The **Rocky Flats Environmental Technology Site** occupies about 140 hectares (~350 acres) near Denver, Colorado, and has more than 400 manufacturing, chemical processing, laboratory, and support facilities that were used to produce nuclear weapons components. Production activities once included metalworking, fabrication and component assembly, and plutonium recovery and purification. Operations at the site ceased in 1989. **Estimated closure date: 2006. Estimated life-cycle costs: $8 billion.**

5. The **Oak Ridge Reservation** covers an area of approximately 155 square kilometers (60 square miles) west of Knoxville, Tennessee. The reservation has three major operating facilities: the Oak Ridge National Laboratory, the Y-12 Plant, and the K-25 Plant. **Estimated closure date: 2014. Estimated life-cycle costs: $6.5 billion.**

6. The **Nevada Test Site**, which occupies about 3,500 square kilometers (1,350 square miles) in southern Nevada, was the primary location for atmospheric and underground testing of the nation's nuclear weapons starting in 1951. **Estimated closure date: 2014. Estimated life-cycle costs: $2 billion.**

agement responsibilities are expected to continue for many decades at an estimated total life-cycle cost of more than $85 billion (DOE, 2000g).

## Disposing of DOE Wastes, Spent Nuclear Fuels, and Nuclear Materials

DOE's Office of Civilian Radioactive Waste Management (RW) is responsible for developing and managing a system to permanently dispose of a currently estimated 85,000 metric tons of heavy metal (MTHM) of commercial spent fuel, 2,500 MTHM of DOE spent fuel, and 22,000 canisters of high-level waste. RW currently is investigating the suitability of the Yucca Mountain Site as a geological repository for such wastes (see Sidebar 2.2). The total cost of disposing of high-level waste, spent nuclear fuel, and nuclear materials in Yucca Mountain is estimated to be $52 to $57 billion over at least the next three to four decades (DOE, 2000g).

EM is responsible for disposing of approximately 167,000 cubic meters (5.9 million cubic feet) of transuranic waste in the Waste Isolation Pilot Plant (WIPP) in Carlsbad, New Mexico (see Sidebar 2.3). The estimated life-cycle costs of WIPP through its estimated closure date of 2039 is $8 billion (DOE, 2000g).

DOE low-level waste is disposed of in shallow land facilities at several locations. Low-level waste from defense programs is disposed of generally at the site where it was produced, primarily Hanford, the Savannah River Site, and the Idaho National Engineering and Environmental Laboratory. Envirocare in Utah receives very low level waste, for example, from facility decommissioning.

## DOE's EQ Challenges

The preceding discussion makes clear that the EQ business line is responsible for managing and controlling a large number of facilities and huge volumes of DOE wastes and contaminated media under a broad range of conditions. For brevity, the committee has developed the following summary statement of the scientific and technical challenges that face the EQ business line, hereafter referred to as DOE's "EQ challenges":

• Remediate (i.e., "clean up") DOE sites and facilities that have severe radioactive and hazardous waste contamination from past activities. In many cases the extent, location, or types of contamination are not

**SIDEBAR 2.2 Yucca Mountain: A Candidate Site for a Geological Repository for High-Level Waste and Spent Nuclear Fuel**

Current U.S. plans call for commercial spent nuclear fuel, high-level waste and spent nuclear fuel at DOE sites, and some nuclear materials (e.g., excess weapons-grade plutonium) to be disposed of in a geological repository. One site, located in Yucca Mountain, Nevada, is being characterized to determine its suitability to serve as a repository.

The 1982 Nuclear Waste Policy Act established a process for siting, developing, licensing, and constructing a geological repository and established an Office of Civilian Radioactive Waste Management within DOE to manage this process. The act also placed primary responsibility for spent nuclear fuel storage on its producers—DOE defense sites and commercial nuclear power plants. DOE initially identified several potential repository sites, including Yucca Mountain; Deaf Smith County, Texas; and Hanford, Washington. The 1987 Nuclear Waste Policy Amendments Act limited site characterization to the Yucca Mountain Site.

DOE completed a viability assessment of its Yucca Mountain Site in 1998. According to the current schedule for the Yucca Mountain Characterization Program, DOE plans to submit a site recommendation to the President in 2001. The President may then submit a site recommendation to Congress, at which point the Governor or the state legislature of Nevada has the right to file a notice of disapproval, which could be overridden by majority votes in both Houses of Congress. If the site is found to be viable and Congress appropriates the necessary funds, DOE plans to submit a license application to the Nuclear Regulatory Commission in 2002. If approved and licensed, Yucca Mountain could begin receiving waste as early as 2010. Current plans call for the repository to remain open for at least 50 years and possibly as long as 300 years before a decision is made to decommission and close the facility. During this preclosure period, performance confirmation and monitoring activities would continue.

well known, and methods to clean them up safely, timely, effectively, and economically are not available. Indeed, in many cases, DOE is unable to defensibly determine whether cleanup is required and its relative priority.

• Manage, stabilize, process, and dispose of a legacy of widely varying and often poorly understood DOE wastes (including spent nuclear fuels and nuclear materials treated as waste) that are potential threats to health, safety, and the environment. The techniques required to characterize, process, and treat these wastes are often undeveloped or poorly realized.

• Provide effective long-term stewardship of DOE sites that have been remediated as well as currently practical but that have residual

---

**SIDEBAR 2.3 The Waste Isolation Pilot Plant**

The Waste Isolation Pilot Plant (WIPP) is the world's first specially constructed deep geologic repository for long-lived radioactive waste. WIPP has sufficient planned capacity to accommodate the entire inventory of U.S. defense transuranic waste (primarily contaminated clothing, tools, equipment, and debris resulting from the manufacture of nuclear weapons and cleanup of weapons production sites). WIPP is located in the semiarid desert of southeastern New Mexico. The repository itself consists of mined shafts, tunnels, and waste disposal rooms in 250-million-year-old bedded salt about 650 meters beneath the land surface.

WIPP opened in 1999 and is currently receiving a few shipments of waste per week from several weapons production sites. The waste is being shipped to WIPP in boxes and 55-gallon drums for direct emplacement in the repository. Once the repository is filled with waste, the access tunnels and shafts will be backfilled to the surface and permanently sealed. The repository is expected to remain open until about 2039.

---

risks to health, safety, and the environment.

• Develop, open, and operate unique, first-of-a-kind facilities for permanent disposal of radioactive spent fuels and high-level wastes, many of which will be hazardous for thousands to hundreds of thousands of years.

• Limit contamination and materials management problems, including the generation of wastes and contaminated media, in ongoing and future DOE operations.

These EQ challenges drive the EQ R&D portfolio.

## EQ BUDGET AND R&D FUNDING

EQ is DOE's second most expensive business line, accounting for $6.7 billion (or 34 percent) of the $19.7 billion DOE budget for fiscal year 2001 (see Figure 2.1a). In contrast, the annual investment in EQ R&D is the smallest of DOE's four programmatic business lines. For fiscal year 2001, funding for EQ R&D was approximately $298 million (or 4 percent) of DOE's total R&D spending (see Figure 2.1b). These budget data suggest that decision makers in DOE, the Office of Management and Budget, and Congress have not viewed R&D as an effective way to meet DOE's EQ responsibilities. One reason for this view may be the incorrect perception that DOE's EQ problems[3] largely will be addressed in the next

---

[3] The term "EQ problems" refers to the set of technical problems that collectively make up the "EQ challenges" described in the text. This is a useful concept in planning an R&D

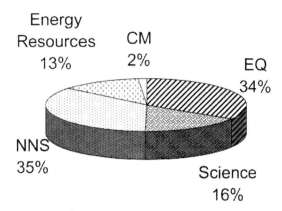

FIGURE 2.1(a) DOE Fiscal Year 2001 Budget by Business Line. Of DOE's total budget of approximately $19.7 billion, $7.0 billion (35%) is spent by National Nuclear Security (NNS), $6.7 billion (34%) by Environmental Quality (EQ), $3.2 billion (16%) by Science, $2.5 billion (13%) by Energy Resources, and 0.3 billion (2%) by Corporate Management and Other (CM). Approximately 41% of DOE's $19.7 billion budget ($8.0 billion) is spent on R&D, which is distributed among the business lines as shown in (b). Data from Department of Energy Office of Chief Financial Officer. Available at:
(http://www.cfo.doe.gov/budget/02budget/3-pager.pdf)

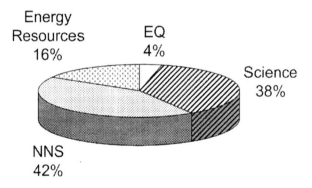

FIGURE 2.1 (b) DOE Fiscal Year 2001 R&D Spending by Business Line. Of DOE's $8.0 billion R&D investment, $3.4 billion (42%) is spent by NNS, $3.0 billion (38%) by Science line, 1.3 billion (16%) by Energy Resources, and $298 million (4%) by EQ. Data for NNS, Science, and Energy Resources are from AAAS (2001); data for EQ are from Ker-Chi Chang, DOE (personal communication).

---

portfolio because the challenges are very broad, and must be broken down into manageable parts to be addressed by R&D.

few years (this issue is discussed below in "Temporal Breadth of DOE's EQ Mission"). The committee examines EQ R&D budget issues more fully in Chapter 5.

## SCOPE OF DOE'S EQ MISSION

The scope of DOE's EQ mission was the subject of extensive deliberation and discussion within the committee. It is important consider this issue early in the report because any consideration of the adequacy of an R&D portfolio requires a clear understanding of the programmatic objectives that these R&D activities are intended to support—in this case, DOE's EQ mission. Such clarity is a challenge because DOE's use of the term "environmental quality" is a misnomer that creates confusion, both within and outside DOE, and because DOE documents reviewed by the committee are not consistent in describing the EQ mission.[4] For example, the EQ strategic goal and objectives in DOE's 1997 Strategic Plan (DOE, 1997b), which was in effect when the EQ R&D portfolio document was compiled, differ substantively in several ways from those in DOE's 2000 Strategic Plan (see Sidebar 2.4). In particular, the 1997 Strategic Plan explicitly recognizes the importance of limiting the generation of future DOE wastes by including "minimize future waste generation" as part of the strategic goal and "prevent future pollution" as one of seven objectives. In addition, the 1997 Plan emphasizes the importance of focusing on the most serious risks (objective 1) and reducing the costs of environmental cleanup (objective 6). None of these important concepts were included in the strategic goal and objectives of DOE's 2000 strategic plan. In spite of these substantive differences, however, the strategic goal and objectives in both strategic plans (see Sidebar 2.4) clearly focus on addressing problems related to DOE wastes and contaminated media.

However, other parts of the 1997 strategic plan and the EQ R&D portfolio document (DOE, 2000b) suggest that the scope of DOE's EQ mission may extend beyond DOE wastes and contaminated media. For example, the 1997 strategic plan states that one of the three primary areas of responsibility of the EQ business line is to "provide the technolo-

---

[4] The committee believes that the most appropriate source for understanding what DOE means by its EQ mission is DOE's published strategic plans. These plans include "strategic goals" and "objectives" for its EQ business line, which together define DOE's EQ mission.

**SIDEBAR 2.4 EQ Strategic Goal and Objectives from DOE's 1997 and 2000 Strategic Plans**

The committee was tasked to identify criteria that should be used to evaluate the adequacy of the DOE's EQ R&D portfolio "in the context of EQ strategic goals and mission objectives." As discussed in the text, the EQ strategic goal and objectives in DOE's 2000 strategic plan differ substantively in a number of ways from those in DOE's 1997 strategic plan, which was in effect when the EQ R&D portfolio document was published. For example, the 1997 strategic plan explicitly recognizes the need to "minimize future waste generation," "reduce the most serious risks...first," and "reduce the life-cycle costs of environmental cleanup"—concepts that are missing from the 2000 strategic plan. Despite these and other differences, however, the strategic goal and objectives from both plans clearly focus on problems associated with DOE wastes and contaminated media, and not on the broad interpretation of DOE's "environmental quality" mission.

*Providing America with Energy Security, National Security, Environmental Quality, Science Leadership.* **U.S. Department of Energy Strategic Plan. September 1997 (DOE, 1997b).**

**EQ Strategic Goal:** Aggressively clean up the environmental legacy of nuclear weapons and civilian nuclear research and development programs, minimize future waste generation, safely manage nuclear materials, and permanently dispose of the Nation's radioactive waste.

**Objective (1)** Reduce the most serious risks from the environmental legacy of the U.S. nuclear weapons complex first.
**Objective (2)** Clean up as many as possible of the Department's remaining 83 contaminated geographic sites by 2006.
**Objective (3)** Safely and expeditiously dispose of waste generated by nuclear weapons and civilian nuclear research and development programs and make defense high-level radioactive wastes disposal-ready.
**Objective (4)** Prevent future pollution
**Objective (5)** Dispose of high-level waste and spent nuclear fuel in accordance with the Nuclear Waste Policy Act as amended.
**Objective (6)** Reduce the life-cycle costs of environmental cleanup.
**Objective (7)** Maximize the beneficial reuse of land and effectively control risks from residual contamination.

*************************************************************************************

*Strength Through Science: Powering the 21st Century.* **U.S. Department of Energy Strategic Plan. September, 2000 (DOE, 2000f).**

**EQ Strategic Goal:** Aggressively clean up the environmental legacy of

*continues on next page*

nuclear weapons and civilian nuclear research and development programs at the Department's remaining sites, safely manage nuclear materials and spent nuclear fuel, and permanently dispose of the Nation's radioactive wastes.

This strategic goal is supported by three objectives:

**Objective (1)** Safely and expeditiously clean up sites across the country where DOE conducted nuclear weapons research, production, and testing, or where DOE conducted nuclear energy and basic science research. After completion of cleanup, continue stewardship activities to ensure that human health and the environment are protected.

**Objective (2)** Complete the characterization of the Yucca Mountain site and, assuming it is determined suitable as a repository and the President and Congress approve, obtain requisite licenses, construct and, in fiscal year 2010, begin acceptance of spent nuclear fuel and high-level radioactive wastes at the repository.

**Objective (3)** Manage the material and facility legacies associated with the Department's uranium enrichment and civilian nuclear power development activities.

gies and institutions to solve domestic and international environmental problems" (DOE, 1997b, p. 24). Taken literally, this would imply that DOE's EQ mission encompasses a wide spectrum of environmental issues (e.g., climate change, biodiversity, ecosystem protection), which as discussed above, is not consistent with the EQ strategic goal and objec-objectives in the same document. Similarly, the EQ R&D portfolio document includes a vision of a significantly expanded future EQ R&D portfolio (termed a "Strategic Portfolio for the 21st Century"), which would include additional R&D investments in areas such as sustainable development and global environmental protection (DOE, 2000b, p. 45). Furthermore, in describing the role of DOE EQ, the document includes the statement that it "provides global leadership to environmental quality efforts" (DOE, 2000b, p. xv). Parts of the 1997 strategic plan and the EQ R&D portfolio document therefore suggest that DOE's EQ mission includes, or should be broadened to include, a myriad of environmental issues beyond DOE wastes and contaminated media.

Given these major inconsistencies, what is the appropriate scope of DOE's EQ mission? The committee discusses three aspects of this issue. The first is the topical breadth of the EQ mission within DOE. In particular, whether DOE's EQ mission includes (or should be broadened to include) environmental issues within DOE beyond wastes and contaminated media. The second is the temporal breadth of DOE's EQ mission, i.e., whether DOE's EQ mission should focus more extensively on long-term problems or in preventing the occurrence of future problems, rather than simply addressing past problems. The third is the national and international breadth of DOE's EQ mission, i.e., whether DOE's EQ responsibilities should be extended to problems outside DOE, such as those in other agencies or nations. This issue is addressed at the end of Chapter 3 (see "Extending the EQ R&D Portfolio Beyond DOE").

## Topical Breadth of DOE's EQ Mission

As discussed above, recent DOE documents have not been consistent in describing the topical breadth of DOE's EQ mission. The EQ strategic goal and objectives in DOE's two most recent strategic plans are quite clear that the topical breadth of DOE's EQ mission is restricted to problems directly related to DOE wastes and contaminated media. However, other parts of the 1997 strategic plan and some parts of the EQ R&D portfolio document suggest that the topical breadth of DOE's EQ mission should be much broader.

Faced with these two very different views, the committee concludes that the more narrow interpretation is more appropriate at the present time. Its reasoning is simple. First, the committee's task explicitly directed the committee to conduct its analysis "in the context of EQ strategic goals and mission objectives," which as discussed above, are quite clear about the topical breadth of the EQ mission. Second, as discussed in Chapter 3, there currently exist a large number of important R&D gaps and opportunities in the EQ R&D portfolio, even within the narrower EQ mission. The committee believes that it would be inappropriate to consider expanding the topical breadth of DOE's EQ mission until the R&D portfolio adequately addresses its current mission. Third, expanding the topical breadth of the EQ mission to include all areas of the environment, such as sustainable development and global environmental protection would create significant overlap with DOE's other missions (in particular, the Energy Resources and Science missions), as well as the missions of other federal agencies with longstanding environmental responsibilities, such as the Environmental Protection Agency, the National Oceanographic and Atmospheric Administration, and the U.S. Geological Survey. Furthermore, such an expansion would make the committee's task nearly impossible, and well beyond the committee's collective expertise.[5]

**Conclusion: The EQ mission should continue to focus on problems associated with DOE wastes and contaminated media.**

---

[5] Readers interested in broader environmental R&D needs are encouraged to read a recent NRC report, *Grand Challenges in Environmental Sciences* (NRC, 2000j).

This conclusion does not exclude the possibility that an expanded mission might be warranted some time in the future when the EQ R&D portfolio adequately addresses the important long-term problems that are already within the EQ mission. It also does not lessen the importance of closely coordinating EQ R&D with related R&D efforts by DOE's other business lines. One of the committee's important conclusions in Chapter 3 is that EQ R&D should build upon the R&D activities of, and take into account the needs of, DOE's Science, Energy Resources, and National Nuclear Security business lines.

Finally, the committee believes that some of the inconsistencies described above arise from the fact that DOE uses the term "environmental quality" as the name of its EQ business line. The term environmental quality is used by many federal and state agencies to refer to a much broader spectrum of environmental issues than problems associated with wastes and contaminated media. For example, issues as diverse as climate change, drinking water protection, ecosystem biodiversity, and protection of marine fisheries would be considered part of environmental quality by agencies such as the Environmental Protection Agency, the National Oceanographic and Atmospheric Administration, and the U.S. Geological Survey. The use of the term "environmental quality" as the name of DOE's EQ business line therefore does not reflect its current mission. The committee encourages DOE to change the name of the EQ business line (and its corresponding R&D portfolio) to more accurately reflect the topical breadth of its EQ mission.

## Temporal Breadth of DOE's EQ Mission

As discussed earlier, the EQ strategic goal and objectives in DOE's 1997 and 2000 strategic plans are not consistent about the importance of addressing long-term problems or in preventing the occurrence of future problems. Both the 1997 strategic plan (objective 7) and 2000 strategic plan (second half of objective 1) recognize a continuing responsibility in the area of long-term stewardship. The 1997 strategic plan also recognizes the importance of minimizing the generation of new wastes. This concept was not included in the 2000 strategic plan, however. The committee believes that it is important for DOE to strive to prevent the types of contamination and waste management problems that have characterized past DOE activities in ongoing and future DOE operations and facilities by assuring that sufficient environmental considerations and protections are built into them up front.[6]

---

[6] The Strategic Laboratory Council's adequacy analysis (DOE, 2000g) recognized the importance of reducing the future waste generation when it recommended a new EQ objective to "minimize the risk, volume and cost of newly generated DOE radioactive and hazardous waste."

Although the inclusion of long-term stewardship as an explicit objective of DOE's EQ mission does recognize an important element of DOE's enduring EQ mission, the committee believes that DOE's strategic plans (especially the 2000 strategic plan) still present a rather limited, short-term view of DOE's long-term EQ responsibilities. This short-term view is reflected in DOE's approach to addressing EQ problems. For example, the recent focus of EM has been to meet the ambitious cleanup goals of the 2006 remediation deadlines and legal or regulatory mandates, such as site implementation plans (DOE, 1998). It is important to recognize, however, that EM's definitions of cleanup for the great majority of DOE sites has meant "securing sites" and "minimizing exposures"—but not rendering the sites suitable for unrestricted use (see Sidebar 1.2; NRC, 2000a). In reality, radioactive and other wastes will remain at most DOE sites even after achieving the cleanup goals, and over 100 sites will require some form of long-term stewardship to protect human health and the environment after they have been closed (NRC, 2000a; DOE, 1999, 2001b). Furthermore, DOE's most contaminated sites with the largest quantities of wastes and contaminated media (e.g., Hanford, Washington; Idaho National Engineering and Environmental Laboratory; Savannah River, South Carolina) will not achieve closure for decades (DOE, 2000e; see Sidebar 2.1).

Similarly, the focus of RW has been to assess the site suitability for licensing of the proposed Yucca Mountain repository by 2010. This short-term emphasis has enabled RW to develop a technical base for determining whether the Yucca Mountain Site could be a suitable geological repository, but generally has not looked beyond licensing to address environmental science, engineering, and social science issues that will arise during the licensing and operation of the repository. For example, improved understanding of the performance of the waste packages within the geological environment and novel monitoring techniques are needed during the pre-closure period, which will last from decades to more than a century.

Although the short-term focus of EM and RW has provided a means for making progress on some short-term elements of the EQ mission, it also may have been misinterpreted by some decision makers to mean that DOE's EQ mission will be essentially completed by 2006 or 2010, i.e., a "going out of business within the next decade" view of DOE's EQ mission. Here the committee needs to explain what it means by the phrase "going out of business within the next decade," because as indicated above, in some respects a going out of business attitude is appropriate for large parts of the EQ business line, and thus for a proportional part of its supporting R&D portfolio. This is because DOE is responsible for sites and materials that today pose serious risks to health and the environment. Thus DOE must act with urgency to mitigate these risks as soon as possible. DOE must put as many as possible of these risks "out of business", e.g., perhaps mitigating the risks by cleaning up a site, or by isolating nuclear wastes in a well-designed repository. It is in this sense, for these treatable risks, that a going out of business attitude is appropriate; EQ's overall mission is to put itself out of business by addressing past problems, and by anticipating and preventing future problems.

However, at present there exists no adequate technology to address other risks, including some of the worst risks. For these, DOE needs parallel programs of long term stewardship and as-long-term-as-necessary R&D to find solutions. And a sense of urgency with respect to these programs is needed in order to protect public health and the environment. Nevertheless, for these risks the sense of urgency does not imply going out of business at any time in the near future. It rather implies getting on with what can be done now, which here is long-term stewardship and R&D. To recapitulate, because DOE faces many serious risks to health and environment, and because some of the worst risks are now unsolvable, the agency must cultivate a balanced sense of urgency, proceeding with deliberate speed to mitigate those risks it can in the short term, and in parallel to initiate R&D on solutions for the currently intractable problems so that in the long term their risks also are addressed. Where this report refers to "going out of business within the next decade," the committee is referring to a mistaken attitude or belief that all EQ problems will be handled in the relatively near future, and the DOE EQ mission completed at that point.

As one might expect, the short-term focus also has had a major impact on DOE's approach to EQ R&D. One of the most consistent and important findings of two recent analyses of DOE's EQ R&D portfolio is that it lacks a long-term strategic vision (DOE, 2000g,h; see also Appendix C). This issue will be discussed at length in Chapter 3.

**Finding**: A "going out of business within the next decade" view of DOE's mission has served to obscure DOE's long-term EQ responsibilities and has done little to address DOE's most challenging EQ problems.

**Recommendation**: DOE should develop strategic goals and objectives for its EQ business line that explicitly incorporate a more comprehensive, long-term view of its EQ responsibilities. For example, they should emphasize long-term stewardship and the importance of limiting contamination and materials management problems, including the generation of wastes and contaminated media, in ongoing and future DOE operations.

The committee's statement of DOE's "EQ challenges" could be used as the basis for these revised strategic goals and objectives.

# 3

# A LONG-TERM VISION FOR DEPARTMENT OF ENERGY ENVIRONMENTAL QUALITY RESEARCH AND DEVELOPMENT

The U.S. Department of Energy (DOE) has taken a first, important step toward integrating its research and development (R&D) programs through portfolio analysis. In so doing, it has recognized the short-term emphasis of the Environmental Quality (EQ) R&D portfolio, and has requested this study to provide strategic advice on how it could build a more effective EQ R&D portfolio. In this chapter, the committee first discusses the important functions of an EQ R&D portfolio, including the associated national and international contexts. The committee then uses these descriptions and the accompanying findings, conclusions, and recommendations to develop a set of criteria to evaluate the adequacy of the portfolio. Finally, the committee discusses five broad themes that DOE could use as "principal elements" of its EQ R&D portfolio. In sum, this chapter represents the committee's vision of a more effective portfolio of activities that incorporates a more "life-cycle based" (i.e., systematic consideration of the entire expected life-cycle of a technology or facility, from initial design, through operation, to closure and long-term stewardship) approach to DOE's EQ problems and moves beyond the short-term, "going-out-of-business within the next decade" philosophy that has driven DOE's EQ R&D to focus on short-term needs over the last decade.

## IMPORTANT FUNCTIONS OF AN EFFECTIVE EQ R&D PORTFOLIO

An effective long-term R&D portfolio could contribute to DOE's EQ mission in a number of important ways. Effective EQ R&D also should contribute significantly to DOE's other missions. In this section, the committee describes the following functions that are considered essential for an effective, long-term EQ R&D portfolio:

- addressing long-term, currently intractable[1] EQ problems;
- improving performance, reducing risks to human health and the environment, decreasing cost, and advancing schedules;
- advancing more informed EQ decision making;
- making informed decisions on nuclear energy;
- promoting national security;
- helping to bridge the gap between R&D and application;
- supporting research and training in relevant fields of science and engineering;
- leveraging results from DOE's Office of Science; and
- leveraging and supporting relevant R&D programs outside DOE.

### Addressing Long-Term, Currently Intractable EQ Problems

The problems confronting the EQ business line are long-term, both because they involve materials that in some cases remain hazardous for thousands to hundreds of thousands of years and because they pose scientific questions that are so complex and unique that R&D will have to continue for decades to generate their solutions.[2] This uniqueness and complexity demand that the EQ R&D portfolio have a strong, if not dominant, long-term component. As discussed in Chapter 2, this contrasts markedly with the current short-term emphasis of DOE's EQ R&D portfolio. In this section, the committee describes some important long-term EQ problems.

One of the most important long-term EQ challenges is long-term stewardship of legacy waste sites for which cleanup is complete but that have residual risks to human health and the environment. As discussed in Chapter 1, radioactive and other wastes will remain at most sites even after achieving the cleanup goals, and active stewardship activities to protect human health and the environment from hazards will be required for long or indefinite time periods (see Sidebar 3.1; DOE, 1999a). One of the most important problems associated with long-term stewardship is the lack of adequate long-term institutional management capabilities, which will require long-term scientific, technical, and social science R&D (NRC, 2000a). Furthermore, DOE's largest sites will require decades to reach their stated cleanup goals (see Sidebar 2.1). Many of the important EQ problems facing DOE at these sites currently have no acceptable, identified solution and will require sustained R&D efforts well beyond

---

[1] The committee uses the term "currently intractable" to refer to problems for which there are no identified, acceptable solutions but for which long-term R&D could lead to such solutions.

[2] When the expression "long-term R&D" is used in this report, the committee means "long-term" from both of these perspectives.

---

**SIDEBAR 3.1 Long-Term Stewardship of Legacy DOE Sites**

Because of the size and complexity of cleanup operations, the life-cycle costs to close DOE sites is now estimated to be $168-212 billion (estimated costs from FY 1997 to 2070 in constant 1999 dollars), and closure at the largest sites will not be completed until as late as 2050 (DOE, 2000e; NRC, 2000a). The term "closure," however, is a misnomer. Even after DOE sites are closed, 129 of them will require continued, long-term surveillance and maintenance, which will include controlling releases from sites, limiting access, and maintaining public records and site markers (DOE, 1999a, 2001b; NRC, 2000a). DOE's responsibilities, and thus financial burdens, for long-term stewardship of these sites will persist for the indefinite future, as will the potential risks to the public and the environment. DOE estimates that it currently spends approximately $64 million annually on long-term stewardship activities, and these costs will increase to nearly $100 million annually by 2050, when all sites are expected to be closed (DOE, 2001b).

The hazards associated with DOE sites and facilities will not be eliminated at any time in the near future (DOE, 1999a). In many cases, such as closed high-level waste tanks, radioactive waste disposal sites, and test and production facilities, the hazards will persist for many thousands of years. The unprecedented scale and longevity of the problems create a host of challenges, such as maintaining records over thousands of years, maintaining control of, and monitoring sites indefinitely, and developing a process for regularly reevaluating the status of the sites and addressing problems when they occur. In order for the sites to be safeguarded effectively over such long timeframes, new technologies will be required and will need to be continually upgraded as technological advancements are made—particularly when risks and costs can be significantly reduced. If the last two decades are at all indicative, for example, information storage will change dramatically over the next hundred years, let alone the next thousand years. It will be essential that data concerning DOE sites be stored in a fashion, and upgraded as necessary, to ensure that it can be accessed far into the future.

The long-term nature of the hazards at DOE sites also means that technologies, societal values, economics, standards, and politics are likely to evolve substantially before environmental remediation and radioactive materials problems are resolved. It will therefore be necessary to use methods that, to the maximum extent possible, will allow subsequent actions to be taken to further stabilize, remediate, or treat materials to reduce risks to the environment and public. This approach requires substantial investment in EQ R&D to ensure that an enduring program is in place to take advantage of future technological developments.

---

2006 (see Sidebar 3.2). For example, proposed solutions for the treatment of high level waste are still being developed (see Sidebar 3.3). In sum, many of DOE's waste management and disposal problems currently are, and will continue to be, intractable during the active clean-up

---

**SIDEBAR 3.2 Remediating and Monitoring Groundwater Contamination at the Hanford Site**

Approximately 1.2 billion cubic meters of groundwater is contaminated with radioactive, hazardous, and toxic substances under the Hanford Site in eastern Washington (DOE, 1997a). In the near term, concerns have been raised about heavy metals, such as chromium, reaching the Columbia river in sufficient quantities to threaten salmon spawning grounds. Concerns also have been raised that, in the longer term (i.e., more than 100 years), substantial quantities of radioactive substances could reach the river and pose significant risks to the environment and local populations.

At present there is no cost-effective means of remediating such a large volume of groundwater, and methods for limiting underground transport are hardly better. Further R&D in the areas of hydrogeology, treatment and extraction technologies, and monitoring will be essential to protect the environment and local populations in the long term. The National Research Council is currently reviewing the Hanford Site's science and technology program for contamination problems associated with the vadose zone and groundwater. A report recommending ways to improve the technical merit and relevance of this program is expected to be released during the summer of 2001.

---

period. A strong continuing R&D portfolio therefore is essential, and may be more important after cleanup than before.

Similarly, there are significant uncertainties and major technical and social science challenges associated with investigating and developing geological repositories for high-level waste and spent nuclear fuel (NRC, 2001). For example, improved understanding of the performance of the waste packages within the geological environment and novel monitoring techniques are needed during the pre-closure period, which is expected to last from decades to several centuries for Yucca Mountain. Such long-term R&D could help ensure that the repository is operating effectively and could allow the repository design to be refined during the pre-closure period to improve its performance and/or reduce costs. Similarly, long-term R&D could help identify and implement measures to build public confidence in repository performance during the pre-closure period.

In summary, the short-term emphasis of EQ R&D efforts described in Chapter 2 and the declining budget trends discussed in Chapter 5 are fundamentally inconsistent with the long-term nature of the problems the EQ business line must address. DOE is responsible for managing, removing (or isolating), and disposing of uniquely hazardous, chemically complex substances, such as spent nuclear fuel, liquid high-level radioactive wastes, and mixtures of hazardous and radioactive compounds. It is also responsible for remediating a wide range of contaminated media and facilities (e.g., groundwater, soil, and nuclear production facilities).

---

**SIDEBAR 3.3 Development of Treatment and Solidification Methods for High-Level Waste at the Savannah River Site**

Treatment and solidification of high-level radioactive waste derived from the production of plutonium and other special nuclear materials is one of the most challenging problems confronting DOE. The Savannah River site stores 120,000 cubic meters of intensely radioactive and chemically complex high-level waste (DOE, 1997a; NRC 2000f). Development of a treatment process has required research into a one-of-kind system and extensive supporting research and engineering.

DOE spent over a decade developing a process (in-tank precipitation) to remove actinides, strontium, and cesium from the high-level waste salt in order to reduce the number of waste canisters that would be produced and sent to a geologic repository for disposal. Despite years of effort and an expenditure of almost $500 million, however, DOE recently was forced to abandon this approach because it did not work as planned (NRC, 2000f). Prior to this decision, EM's Office of Science and Technology had supported some R&D on alternatives to in-tank precipitation. Although limited, this R&D helped DOE initiate a major R&D effort on alternative technologies to ensure that the waste salt processing could proceed in an efficient and reliable manner.

---

These activities must be carried out under a wide range of challenging and often unique circumstances. In many cases, environmental remediation, management, and disposal of hazardous and radioactive substances require development of innovative technologies. Environmental cleanup, waste management, and disposal activities will, of necessity, endure for generations, and long-term stewardship at most sites could continue indefinitely thereafter. Therefore, the future can provide opportunities for continual improvements in the methods used to address these issues and the possibility of breakthrough technologies that could greatly reduce the risks to human health and the environment and the costs to future generations.

**Finding: Many of the problems confronting the EQ business line are long-term, both because they involve materials that in some cases remain hazardous for thousands to hundreds of thousands of years, and because they are so complex and unique that R&D will have to continue for decades to generate their solutions.**

**Conclusion: The uniqueness and complexity of DOE's EQ problems demand that the EQ R&D portfolio have a strong, if not dominant, long-term component.**

**Recommendation**: DOE should begin to devote an increasing fraction of its EQ R&D to long-term problems to ensure that an R&D portfolio dedicated to long-term problems is in place within five years.

**Conclusion**: The technical and social complexities associated with nuclear materials handling, storage, waste management, and disposal demand a clear long-term vision.

**Recommendation**: DOE should develop a long-term strategic vision for its EQ R&D portfolio. This vision should provide the framework for developing the science and technology necessary to address EQ problems that extend beyond the present emphasis of short-term "compliance" and should incorporate the principle of continual improvement.

The importance of long-term EQ problems does not mean that DOE should focus its EQ R&D efforts exclusively on long-term problems. Short-term R&D should be undertaken to address near-term problems, such as those driven by legal and regulatory requirements (e.g., cleanup of contaminated groundwater, see discussion in NRC, 1999b). It is essential, however, that the anticipated timeframe of such R&D (i.e., when results can be expected) be consistent with the short timeframe of such problems. Long-term R&D should not be undertaken on problems that will be addressed in the near term.

## Improving Performance, Reducing Risks to Human Health and the Environment, Decreasing Cost, and Advancing Schedules

The type of problem-driven R&D envisioned as part of DOE's EQ R&D portfolio should be viewed as an investment (see discussion in Chapter 4). The results should be expected to improve performance, reduce risks to human health or the environment, decrease costs, or advance schedules. Successes and failures should be closely monitored and additional investment made if R&D has paid off well.[3] Failure of past levels of R&D to pay off is an indication that one or more of the following may be true: The portfolio was not balanced, the program was poorly managed, the funding was too high, the wrong researchers were involved, or the evaluation was premature (i.e., taking place before R&D results have been realized). Furthermore, an R&D portfolio that rewards

---

[3] The success or failure of a new technique or method in achieving one or more of these objectives is directly related to whether the R&D results are "deployed" in the field. Deployment is necessary but not sufficient for success, as some deployments may not improve performance, reduce risks, decrease costs, or advance schedules.

innovation in solving current problems that are extremely challenging or unacceptably expensive will have a certain number of marginal successes or outright failures. That is among the signs that an R&D program is healthy and pushing the cutting edge of science and technology. Furthermore, knowledge gained through R&D failures can be very useful. Even so, an important measure of the long-term success of the R&D portfolio is the degree to which it has led to improved performance, reduced risks to human health or the environment, decreased costs, and advanced schedules.

Although these four objectives can be used as a measure of the success of the EQ R&D portfolio, the types and timeframes of R&D need to be considered when identifying appropriate metrics for success of individual projects. Long-term R&D (especially fundamental research) often carries inherently greater risks, can take many years to come to fruition, and can result in benefits in unexpected applications. It would be a mistake to expect all research to lead to demonstrable results in a very short time, or to avoid the risk of failure by excluding R&D to address particularly challenging problems. The success of long-term research projects can be evaluated periodically through peer review (COSEPUP, 1999b; NRC, 1998); whereas the success of more applied R&D projects can be evaluated through relatively direct measurements (COSEPUP, 1999b), such as the development of a new technology that is more effective, less costly, or more time efficient than earlier technologies. Different types of R&D carry with them differing expectations, and it is important in evaluating success to calibrate expectations to the type of work being done.

**Conclusion: Careful analysis of the success and failures of R&D over time is an important consideration in evaluating the adequacy of the EQ R&D portfolio and in determining an appropriate level of EQ R&D investments.**

**Recommendation: DOE should institute a program to analyze periodically the impact of the R&D portfolio and should take into account the success of past R&D investments in making future R&D funding decisions.**

These analyses should not preclude R&D with a significant risk of negative results if the potential gain is substantial. Metrics for the portfolio as a whole should include measurements of the degree to which it has led to improved performance, reduced risks to human health and the environment, decreased costs, and advanced schedules. Metrics for individual projects should reflect the differing objectives and timeframes of various R&D projects, such as fundamental research and applied R&D. Such metrics should be developed with input from independent experts such as the advisory group recommended later in this report (see Chapter 4).

## Advancing More Informed EQ Decision Making

In many cases, the availability of improved information and scientific, technical, and social understanding can lead to more informed decision making. For example, more efficient and cost-effective technologies based on improved technical understanding could reduce the costs of remediating contaminated DOE sites. However, numerous decisions on environmental remediation, waste management, materials storage, and facility decommissioning involve complex technical issues for which there are only limited data and partial scientific understanding. Recent studies have identified major gaps in scientific and technical understanding related to EQ problems, including subsurface science (NRC, 2000c; DOE, 2000g), the complex chemical dynamics in high-level waste (NRC, 2000d; DOE, 2000g), corrosion rates for materials used for long-term storage and disposal of high-level waste (NRC, 2000d; DOE, 2000g), and the mobility of certain heavy metals in surface and groundwater (NRC, 1999b; DOE, 2000g). These knowledge gaps affect DOE's decision-making in a number of important areas, including the following:

- understanding fully the risks to human health and the environment that are associated with DOE wastes and contaminated media;
- determining the magnitudes and types of technical, scientific, and social uncertainties with which DOE programs contend;
- balancing effectively the risk and rewards of various options for cleanup, end states, storage, treatment, and stewardship of hazardous, toxic, or radioactive materials (i.e., life-cycle analyses);
- avoiding or minimizing environmental harm and risks to human health that are associated with meeting national security responsibilities; and
- addressing environmental remediation and long-term stewardship responsibilities associated with existing or future national and international energy needs.

In short, there is an array of issues, ranging from disposal of high-level waste to remediation of environmental contaminants to construction of new research facilities for ongoing defense programs, that could benefit from further EQ R&D underpinning defensible, enduring decision making.

It should be emphasized, however, that lack of technical information does not necessarily preclude effective decision making. Current decisions must consider that technology and understanding can be expected to improve considerably during the long timeframes of some EQ challenges. For residual contamination at closed legacy sites, for example, the system of long-term stewardship put in place should not preclude future actions to address remaining risks to human health and the environment (see Sidebar 3.1). The system should allow future decision makers to re-initiate active cleanup activities if and when future technologies or understanding develop to a point where it makes sense to address remaining risks (NRC, 2000a), or when the understanding of the risks to human health and the environment improves. For geological disposal of high-level wastes and spent nuclear fuel, DOE should pursue a phased approach that would allow changes to the disposal plans to improve operations, safety, and schedule or reduce cost throughout the decades-long process of emplacement (see Sidebar 3.4). Such a phased decision making process[4] also was recommended for dealing with high-level waste problems at the Hanford Site (NRC, 1996b), and could be applied to a number of the most important long-term EQ problems.

In addition to filling science and technology gaps, effective long-term R&D programs also support R&D on technical alternatives when existing techniques are expensive, inefficient, or pose high risks to human health or the environment, or where techniques under development have high technical risks[5] (NRC, 1999a; DOE, 2000g). Several recent studies have found that the EQ business line has not adequately supported such R&D in the past, and have recommended that strategic R&D on technical alternatives be added to the portfolio (NRC, 1999a; DOE, 2000g). When information is inadequate to make the decision desired, i.e., to choose between major policy options, one can seek more information in two ways. The two paths can be taken in parallel or as alternatives, depending on the policy situation. One is to initiate R&D (perhaps postponing the decision). Global climate change illustrates this option. The second is to take a more modest decision that may yield more information (i.e., "experience") and which leaves open the major policy options. This latter approach does not preclude initiating R&D in parallel.

**Finding: Numerous decisions on environmental cleanup, waste management, materials storage, and facility decommissioning involve complex technical issues for which only limited data and partial scientific understanding exist.**

**Conclusion: The EQ R&D portfolio is critical to improving decision making and should be designed to help inform important DOE decisions, including support for technical alternatives in areas of high cost or high risk.**

---

[4] Also commonly referred to as "adaptive management."
[5] Technical risk is defined as "the probability that the technique or method fails to accomplish the goals and performance requirements set by policy or regulation."

---

**SIDEBAR 3.4 A Flexible Approach to the Disposal of High-Level Radioactive Waste and Spent Nuclear Fuel**

One of the most difficult of DOE's EQ challenges is the need to develop, open, and operate unique, first-of-a-kind facilities for the permanent disposal of radioactive spent fuels and high-level wastes. In 1990, the National Research Council's Board on Radioactive Waste Management published a report, *Rethinking High-Level Radioactive Waste Disposal*, which suggested that DOE adopt a flexible and experimental institutional approach to this challenge. In particular, the report described a strategy that acknowledges the following premises:

- *Surprises are inevitable in the course of investigating any proposed site, and things are bound to go wrong on a minor scale in the development of a repository.*
- *If the repository design can be changed in response to new information, minor problems can be fixed without affecting safety, and major problems, if any appear, can be remedied before damage is done to the environment or to public health.*

*This flexible approach can be summarized in three principles:*

- *Start with the simplest description of what is known, so that the largest and most significant uncertainties can be identified early in the program and given priority attention.*
- *Meet problems as they emerge, instead of trying to anticipate in advance all the complexities of a natural geological environment.*
- *Define the goal broadly in ultimate performance terms, rather than immediate requirements, so that increased knowledge can be incorporated in the design at a specific site.*

*In short, this approach uses a scientific approach and employs modeling tools to identify areas where more information is needed, rather than to justify decisions that have already been made on the basis of limited knowledge.* (NRC, 1990, p. 7)

---

### Making Informed Decisions on Nuclear Energy

Today it is not clear how and by which technologies the current problems facing nuclear energy may be resolved. What actually happens will depend on how safety, waste disposal, and proliferation concerns are resolved, and whether the greenhouse debate adds increasing importance to nuclear energy's "carbon benignness." (IIASA, 1995, p. 62).

**power sources in order to integrate EQ R&D with relevant Energy Resources programs.**

## Promoting National Security

With the end of the Cold War, national security and non-proliferation objectives led the United States to weapons dismantlement and cessation of weapons testing. This gave rise to dramatically new nuclear materials stewardship responsibilities for the United States and other countries, especially Russia. These new responsibilities, in turn, have changed U.S. needs regarding the storage, processing, possible uses, waste management, and disposal of excess special nuclear materials. In response, DOE established the National Nuclear Security business line, which is designed to enhance national security through the military application of nuclear technology and reduce the global danger from weapons of mass destruction (DOE, 2000f).

There are a number of potential benefits from a successful EQ R&D effort that can have major impacts on the success of DOE's national security mission. The National Nuclear Security business line has six objectives, including two that are directly relevant to the EQ business line:

• reduce the global danger from the proliferation of weapons of mass destruction; and
• ensure that DOE's nuclear weapons, materials, facilities, and information assets are secure through effective safeguards and security policy, implementation, and oversight.

Successfully meeting these goals will require close collaboration with the EQ business line. Reducing nuclear weapon stockpiles requires facilities and operations that could create new environmental problems, and the environmental and waste management aspects should be considered up front in program decisions. For example, an option that minimizes wastes or results in wastes that are in a better form for disposal might be preferable to one that is perhaps a little cheaper but leaves a bigger waste management or facility cleanup problem. These nuclear materials, regardless of their origin, will need to be managed, processed, stored, transported, and ultimately disposed of permanently. National security interests are directly affected by, for example, EQ R&D on processes that could be used to dispose of surplus plutonium by immobilizing it for disposal in vitrified high-level waste (DOE, 2000c).

Without effective EQ R&D, the disposition of materials arising from the dismantlement of nuclear weapons could be impeded, undermining national security. This issue illustrates the importance of effective linkages among different DOE program units (Offices of Environmental

Management [EM]; Civilian Radioactive Waste Management; Nuclear Energy, Science and Technology; and the Office of Fissile Materials Disposition) and R&D portfolios (EQ, National Nuclear Security, and Energy Resources). The success of DOE's national security mission, therefore, is dependent upon DOE's EQ mission, which requires that the latter have an effective long-term R&D program.

**Finding: There are a number of potential benefits from a successful EQ R&D effort that could have a significant impact on U.S. national security.**

**Conclusion: DOE's R&D planning efforts should consider the value of EQ R&D to DOE's national security mission and the potential impacts on EQ R&D requirements arising from national security mission decisions.**

## Helping to Bridge the Gap Between R&D and Application

Outside reviews have found that information and technologies developed by the EQ R&D portfolio often are not promptly used in the field by DOE contractors. In a 1997 NRC report on DOE's Environmental Management Science Program (EMSP), for example, the committee found that the movement of new knowledge and insights from investigators to full-scale application is a slow and diffuse process (NRC, 1997). A number of reports from the U.S. General Accounting Office have discussed problems that have hindered the movement of technologies developed and demonstrated within EM's Office of Science and Technology into the field (GAO, 1996, 1998). The problem of achieving effective implementation of new technologies is not unique to DOE (e.g., NRC [2000b] discusses this problem in the area of weather prediction), and its causes are numerous and widespread.

Several factors specific to DOE exacerbate the problem of deploying new or novel technologies.

1. At many sites cleanup proceeds under operational contracts that do not provide any incentive for contractors to adopt new technologies— in fact, contractors' incentives often run counter to adopting new technologies that accelerate a project, because the longer a contract lasts the more it is worth (NRC, 1999b).

2. Legal or regulatory requirements may, for good reasons, specify a certain technological approach or timetable for cleanup actions. Prescribing a particular technology or schedule, however, can effectively foreclose innovation.

3.  Political pressures may prompt or prevent deployment of technologies—regardless of whether they make sense technically.

4.  New technologies may remain unused because site managers prefer well-established technologies, because they are familiar with them and because they are unwilling to accept the higher risk of violating legally mandated schedules attendant with adopting novel approaches.

5.  Technology transfer is frequently impeded by weak feedback channels between EQ R&D and operational personnel (DOE, 2001a). For example, one of the principal channels used in EM, the Site Technology Coordination Groups,[7] have been criticized as being overly formal, complex, cumbersome and slow, and focused more on developing new projects than promoting effective coordination and deployment (NRC, 1999a).

These factors have a common theme: Constraints beyond and distinct from the specific EQ problem being addressed override the motivation to deploy technologies derived from EQ R&D. In some instances, the causes are outside the control of EQ R&D managers (e.g., contractual requirements); in others, R&D managers could act to reduce the disconnect that arises between R&D and deployment (e.g., improving contacts with future users to ensure deployments). In either case, EQ R&D must remain focused on the real problems, while at the same time DOE management must develop effective mechanisms to eliminate or at least buffer the EQ R&D portfolio against systemic impediments such as contractor parochialism and contractual disincentives.

The discussion above touches on many reasons why R&D often is not applied in the field. The particulars vary from site to site and case to case. Accordingly, the remedies may be many and varied. One potential approach to this problem would be to explore a variety of remedies on an experimental (e.g., limited and reversible) basis at a variety of sites. The success of each would be monitored and evaluated, and results documented and disseminated. In this way a set of validated approaches could be developed which could be modified for local situations and adopted as appropriate. Such an approach builds on local experience to avoid the common "one size fits all" failure.

**Finding: Information and technologies developed in the EQ R&D portfolio often are not promptly used in the field by DOE contractors.**

---

[7] EM's Office of Science and Technology formed Site Technology Coordination Groups at each major site to interact with local contractor personnel and others to obtain that site's environmental restoration and waste management technology needs.

**Conclusion**: The gap between R&D and application has many causes, some of which can be addressed by EQ R&D managers while others are outside their control.

**Recommendation**: DOE's EQ R&D managers should provide continual feedback to users (and accept input from users) and include sufficient funds and incentives to improve application of R&D results where they will solve EQ problems.

### Supporting Research and Training in Relevant Fields of Science and Engineering

A strong EQ R&D portfolio requires technically skilled individuals. Research and training programs in nuclear engineering, radiochemistry, and related fields of science and engineering have been decreased substantially in recent decades (DOE, 2000k). An effective, adequately funded R&D portfolio that includes new starts, extensions of promising R&D, and periodic new initiatives has the potential to promote the development of the future nuclear and environmental scientists and engineers required to address the long-term problems described in this report. For example, undergraduate and graduate students with an interest in environmental and nuclear fields (and their advisors) must view the EQ R&D portfolio as providing sustained support for "cutting edge" R&D to address important national problems. The portfolio needs to attract and retain a cadre of top-quality researchers in academia and the national laboratories who are knowledgeable and committed to DOE's R&D needs, and help support the students, postdoctoral associates, and faculty necessary for the enduring mission. In addition, DOE needs to help develop people with practical training in the handling of hazardous materials, operation of facilities containing such materials, packaging, and transportation.

As discussed earlier, DOE's EQ strategic objectives have not been consistently and clearly articulated in high-level DOE planning documents (see discussion in Chapter 2). To attract and retain top-tier scientific and engineering talent, the R&D portfolio must have a clear vision and stable funding. Enhancing the stability of funding could be achieved by a variety of means, such as ensuring that new funding cycles occur on a regular, if not annual, basis and by making longer-term grants available to researchers. Issues associated with R&D funding levels are discussed more fully in Chapter 5.

**Finding**: Research and training programs in relevant fields of science and engineering have been substantially reduced in recent decades.

**Conclusion**: An effective and adequately funded EQ R&D portfolio that includes new starts, extensions of promising R&D, and periodic new initiatives could promote the development of the future scientists and engineers required to address DOE's long-term EQ problems.

**Recommendation**: The EQ R&D portfolio should include stable support for research and training in relevant fields of science and engineering, including periodic new initiatives on important EQ problems.

The R&D centers recommended in Chapter 4 would be a good way to develop such people. It should be noted, however, that the EQ R&D portfolio cannot be expected to provide all of the support that is necessary to develop the future scientists and engineers to address DOE's long-term EQ problems. One of the objectives of DOE's Science business line is to provide the "scientific workforce...that ensures success of DOE's science mission and supports our Nation's leadership in the physical, biological, environmental, and computational sciences" (DOE, 2000f, p. 7). Therefore, DOE's Office of Science also could be expected to help meet these needs. Other important sources of federal support for research and training in relevant areas of science and engineering include the National Science Foundation, the Environmental Protection Agency (EPA), and the Department of Defense.

### Leveraging Results from DOE's Office of Science

DOE's Science business line is dedicated to "advanc[ing] the basic research and instruments of science that are the foundations for DOE's applied missions, a base for U.S. technology innovation, and a source of remarkable insights into our physical and biological world and the nature of matter and energy" (DOE, 2000f, p. 7). The Science business line funds basic research in four central areas: (1) fueling the future (clean and efficient energy sources), (2) protecting our living planet (environmental impacts of energy production); (3) exploring matter and energy, and (4) extraordinary tools for extraordinary science (e.g., multidisciplinary research).

DOE's 2000 strategic plan also directs the Office of Science (SC) to "support long-term environmental cleanup and management at DOE sites...." (DOE, 2000f, p. 7). SC includes a number of basic research programs that are related to DOE's EQ mission (see Table 3.1), particularly in its Office of Basic Energy Sciences (BES) and the Office of Biological and Environmental Research (BER). Among other programs, the BES supports projects to improve current understanding and to mitigate the environmental impacts of energy production (DOE, 2000b). Areas of particular importance to the EQ R&D portfolio include the geoscience, separations science, and materials science and engineering research programs. The BER focuses on research designed to advance environmental and biomedical knowledge connected to energy (DOE, 2000b); examples of areas of relevance to the EQ R&D portfolio include the Natural and Accelerated Bioremediation Program and the Environmental Molecular Sciences Laboratory.

It is important to recognize that although some SC research is relevant to EQ R&D, the main drivers for most SC research are not EQ problems or the problems addressed by DOE's other mission areas. Rather, SC's research is inherently "basic" (i.e., it looks within science for its research questions and justifications). Put another way, SC sees research as an end in itself, but for EQ research is a means to an end. These different world views make cooperation and coordination difficult, and unlikely without conscious, continual effort. The EMSP program, which has been noted as making important research contributions by a number of recent studies (DOE, 2001a; NRC, 2000c, 2001d,e) demonstrates that such cooperation and coordination are possible.

**Finding**: **Significant elements of DOE's Science portfolio are directly related to components of the EQ R&D portfolio.**

**Conclusion**: **Relevant research supported by DOE's Office of Science should be integrated and coordinated with EQ R&D.**

### Leveraging and Supporting Relevant R&D Programs Outside of DOE

To provide a broader context and as part of its task to consider whether the EQ R&D portfolio should incorporate related issues outside DOE, the committee considered relevant R&D programs in other agencies, the private sector, and other nations. The committee gathered information during its August workshop, through Internet searches and direct communication with program managers, and by reviewing recent studies examining related R&D programs (e.g., NRC, 2000c; DOE, 2000g). Summaries of a number of related R&D programs are provided in Appendix E. This analysis was necessarily limited, and the list of R&D programs in Appendix E and the discussion that follows, should be read with this caveat in mind.

A recent NRC report, *Research Needs in Subsurface Science* (NRC, 2000c), identified 18 federal R&D programs in 8 agencies that support research "closely related" to DOE's EMSP research to solve subsurface contamination problems at its facilities. Thus, just in this one area, there

TABLE 3.1 Related Research Activities Sponsored by DOE's Office of Science

| Research Activity | FY 00 Budget (million $) | FY 01 Budget Request (million $) |
|---|---|---|
| Natural and Accelerated Bioremediation Research Program | 25.2 | 21.1 |
| Cleanup Research | 3.4 | 2.7 |
| Waste Management | - | 8.1 |
| Heavy Element Chemistry | 6.7 | 7.4 |
| Chemical Energy and Chemical Engineering | 9.0 | 10.0 |
| Analytical Chemistry Instrumentation | 4.6 | 5.8 |
| Separations and Analysis | 12.6 | 14.6 |
| Materials Chemistry | 25.8 | 27.6 |
| Mechanical Behavior and Radiation Effects | 16.6 | 16.4 |
| Health Risks from Low Dose Exposures | 18.3 | 11.7 |
| Environmental and Molecular Sciences Laboratory | 28.8 | 32.4 |
| Geosciences | 15.0 | 15.2 |
| Energy Biosciences | 25.0 | 28.0 |

Source: DOE 2000d.

are numerous related programs in other federal agencies. The committee achieved similar results in its broader, admittedly limited, review of R&D programs that address environmental problems closely related to those in DOE's EQ business line. The range of issues addressed and the number of related programs in other federal agencies and abroad is illustrated by the following examples:

• The Strategic Environmental Research and Development Program is the Department of Defense's counterpart to the DOE EQ R&D portfolio, and is operated in conjunction with DOE and EPA, as well as other federal agencies. The program supports, for example, R&D to develop improved approaches and processes to decrease the quantity of disposed wastes; to increase effective waste management efforts; and to decrease life-cycle, safety, and pollution impact costs.

• The Environmental Security Technology Certification Program, also in the Department of Defense, demonstrates and validates promising innovative technologies in the areas of environmental cleanup and compliance, pollution prevention, alternative waste processing technologies, and detection and remediation of unexploded ordnance.

- The U.S. Nuclear Regulatory Commission's Radiation Protection, Environmental Risk and Waste Management Branch develops, plans, and manages research programs related to the movement of radionuclides in the environment and consequent dose and health effects to the public and workers as a result of nuclear power plant operation, facility decommissioning, clean-up of contaminated sites, and disposal of radioactive waste.
- EPA's Waste Research Strategy addresses issues pertaining to the proper management of solid and hazardous wastes and the effective remediation of contaminated media. It focuses on four research areas: (1) contaminated groundwater; (2) contaminated soils and the vadose zone; (3) emissions from waste incinerators; and (4) active waste management facilities.
- EPA's National Risk Management Research Laboratory conducts research on methods to prevent and reduce risks from pollution that threatens human health and the environment. Its projects include evaluating the cost-effectiveness of methods for prevention and control of air, land, and water pollution; remediation of contaminated media; and restoration of ecosystems.
- EPA's Superfund Innovative Technology Evaluation Program focuses on the development of alternative or innovative treatment technologies.
- EM's Office of Science and Technology and EPA's Office of Solid Waste recently signed a memorandum of understanding to improve cooperation on the development of technical solutions and regulations to address environmental problems associated with mixed wastes.
- The Electric Power Research Institute operates a decommissioning technology program designed to assist utilities in minimizing the cost of decommissioning through enhanced planning, application of lessons learned by other utilities with retired plants, and use of advanced technology. Projects include development of technologies for chemical decontamination, site characterization, and concrete decontamination.
- There are a number of R&D programs in other countries, such as Britain, France, and Japan, that focus on disposal of radioactive wastes—particularly high-level radioactive waste (NRC, 2001).
- The National Institute of Health's Superfund Basic Research Program conducts research on the human health and ecological risks of hazardous substances and promotes the development of new, cost-effective environmental technologies.

In short, there are numerous U.S. government R&D programs that are closely related to R&D activities supported by the EQ business line. Areas of significant overlap include remediating contaminated groundwater and sites, reducing waste generation, and understanding the fate and transport of contaminants in the subsurface. For some of these overlapping issues, DOE is involved in cooperative efforts with other agencies, such as the Strategic Environmental Research and Development Program and the memorandum of understanding with EPA on mixed waste issues. Further, some EQ R&D objectives that are addressed by few other domestic R&D programs, such as management, treatment, and disposal of high-level waste, are being actively pursued by parallel programs in other countries. Accordingly, there are significant opportunities for EQ R&D to benefit efforts outside of DOE and even outside of the United States. Similarly, there are many opportunities to leverage the important R&D conducted outside DOE to help address DOE's EQ problems. In areas where DOE's EQ mission directly overlaps with the missions of other agencies, such as EPA and the Department of Defense, DOE should continue to look for opportunities to coordinate its R&D activities with those agencies.

**Finding: A number of programs in federal agencies outside DOE and in other countries support R&D closely related to DOE's EQ mission. Specific areas where there is significant overlap include remediating contaminated media, reducing waste generation, and disposing of waste.**

**Recommendation: DOE should leverage the information and technologies developed in programs outside DOE and, to the extent possible, coordinate its EQ R&D with related R&D efforts in other agencies. It also should make available the information and technologies developed in the EQ R&D portfolio to industry, other federal and state agencies, and other countries.**

## CRITERIA TO EVALUATE THE ADEQUACY OF THE EQ R&D PORTFOLIO

An important part of the committee's task was to develop criteria that could be used to evaluate the adequacy of the EQ R&D portfolio. The committee used its descriptions of the essential functions of an effective EQ R&D portfolio from earlier in this chapter and the accompanying findings, conclusions, and recommendations to develop the following criteria to evaluate the adequacy of the portfolio:

1.  **There should be no significant gaps in critical areas of science and technology that are required to address EQ goals and objectives.**
2.  **The portfolio should support the accomplishment of closely related DOE and national missions.**

3. The portfolio should include R&D to develop technical alternatives in cases where (1) existing techniques are expensive, inefficient, or pose high risks to human health or the environment; or (2) techniques under development have high technical risk.

4. The portfolio should produce results that could transform the understanding, need, and ability to address currently intractable problems and which could lead to breakthrough technologies.

5. The portfolio should leverage R&D conducted by other DOE business lines, the private sector, state and federal agencies, and other nations to address EQ goals and objectives.

6. The portfolio should help narrow and bridge the gap between R&D and application in the field.

7. The portfolio should be successful in improving performance, reducing risks to human health and the environment, decreasing cost, and advancing schedules.

8. There should be an appropriate balance between addressing long-term and short-term issues.

9. A diversity of participants from academia, national laboratories, other federal agencies, and the private sector, including students, postdoctoral associates, and other early-career researchers, should be involved in the R&D.

10. There should be an appropriate balance of annual new starts, extensions of promising R&D, and periodic new initiatives.

Recommendation: DOE should use, at a minimum, these 10 criteria to evaluate the adequacy of its EQ R&D portfolio.

Most of these criteria require expert evaluations, and therefore will not provide a simple "yes" or "no" answer as to the adequacy of the portfolio (the committee discusses a process for obtaining such expert input in Chapter 4). Even so, such criteria provide a framework that decision makers in DOE, the Office of Management and Budget, and Congress could use to set performance goals and measures for the EQ R&D portfolio and to help prioritize funding decisions. The committee also chose to frame the criteria in terms of substantive goals for an effective R&D portfolio, rather than in terms of the resources required to achieve these goals. The criteria can be directly related to methods to determine appropriate investment levels, as discussed in Chapter 5.

## PRINCIPAL ELEMENTS OF AN EFFECTIVE EQ R&D PORTFOLIO

An important part of the committee's task was to provide advice to DOE on the principal elements of its EQ R&D portfolio. The committee approached this task in two ways: (1) by developing its own list of principal elements and (2) by developing a general methodology that DOE could use continually to identify and refine the principal elements of the portfolio. The committee's list of principal elements is presented below, and the general methodology is discussed in Chapter 4.

These principal elements were derived by analyzing the EQ business line's most pressing problems, the existing gaps in its R&D portfolio, the areas presenting the greatest opportunities for improvement, and by applying the criteria discussed in the previous section. One of the most important sources of input was the Strategic Laboratory Council's (SLC's) adequacy analysis of the portfolio (DOE, 2000g), which included a detailed analysis of the R&D gaps and opportunities in the portfolio (see Appendix C). The committee also reviewed a number of recent studies on aspects of the portfolio (see Appendix F for an annotated bibliography of the National Research Council studies that were reviewed); solicited input from experts knowledgeable about DOE's EQ mission; and convened a public workshop in August 2000 to discuss this issue and other aspects of the committee's charge (see Appendix B for workshop agenda and list of participants).

In identifying the principal elements, the committee took a high-level, long-term view of the R&D needed to address DOE's most challenging EQ problems. The committee did not define elements along existing DOE program lines, but attempted to identify crosscutting themes that apply to a number of DOE's program units. The five principal elements therefore are quite broad. To illustrate the crosscutting nature of these elements and to document the committee's basis in recommending these elements, the discussion that follows includes numerous citations to previous studies from the NRC, DOE, and other groups.[8] The topics discussed below do not constitute a comprehensive list of long-term EQ R&D needs; rather, it is intended to provide DOE with a useful starting point from which to build a more effective, long-term R&D portfolio. The committee describes how DOE could build upon this list of program priorities to achieve and maintain a more strategic EQ R&D portfolio in Chapter 4.

**Recommendation**: The EQ R&D portfolio should include, at minimum, the following 5 principal elements:

**1.   Development and evaluation of approaches that reduce the impacts of wastes on human health and the environment through generation minimization; processing improvements, including volume reduction, stabilization, and containment; and disposal;**

---

[8] Although the committee has attempted to briefly synthesize the relevant message from each referenced work, readers interested in more details on any subject are encouraged to read the complete reports, where the rationales for conclusions and recommendations are described.

2.  Development of methods and techniques for cutting-edge characterization and remediation of contaminated media, including facilities;

3.  Improvement in understanding the movement and behavior of contaminants through the environment, with an emphasis on locating and tracking the movement of contaminants in the subsurface;

4.  Development of mechanisms for effective long-term stewardship, including improved institutional management capabilities, appropriate monitoring, and the means to implement future improvements in technology and understanding; and

5.  Determination of the risks of DOE wastes and contaminated media to human health and the environment to improve the bases upon which regulatory and societal decisions can be made.

Each of these principal elements is described in more detail in the sections that follow.

## Development and Evaluation of Approaches that Reduce the Impacts of Wastes on Human Health and the Environment

Recent studies have identified significant long-term R&D needs in three general areas related to reducing the effects of DOE radioactive, hazardous, and mixed wastes to human health and the environment: (1) generation minimization; (2) waste processing improvements, including volume reduction, stabilization, and containment; and (3) disposal.

### Generation Minimization

Minimizing the generation of DOE wastes (both new wastes and secondary wastes produced during remediation activities) is an essential element of the life-cycle approach to EQ problems emphasized throughout this report. As discussed previously, reducing the environmental consequences of future nuclear power technologies (both wastes and contaminated facilities) has been recognized as an important long-term R&D need (DOE, 2000k). Moreover, a recent NRC committee and DOE's adequacy analysis both identified the minimization of the generation of new wastes as an important and promising area for DOE R&D (NRC, 2000a; DOE, 2000g). DOE's EQ R&D portfolio does not include a specific program for waste minimization, although EM does support some R&D projects related to generation minimization during cleanup activities, and DOE has included some department-wide efforts to reduce pollution and waste in response to the Greening the Government Executive Orders (DOE, 2000g; White House, 1999).

## Waste Processing Improvements

A recent NRC committee and the SLC's adequacy analysis both concluded that long-term R&D is needed to develop high-efficiency separation methods for high-level waste in order to minimize the environmental impacts of secondary wastes generated during its processing (NRC, 2000d; DOE, 2000g). Recent studies also have highlighted the importance of long-term research on new waste containment and stabilization technologies, particularly for high-level waste (NRC, 1999c, 2000c,d; DOE, 2000g). Long-term stabilization and containment is also a critical component of DOE's long-term stewardship responsibilities (NRC, 2000a; DOE, 2000g). Other reports have noted the need for long-term R&D to address the lack of final waste acceptance criteria for high-level wastes (NRC, 2000g; DOE, 2000g).

## Disposal

With regard to the disposal of high-level waste and spent nuclear fuel, a December 2000 letter report from the Nuclear Waste Technical Review Board identified the conceptual design for the proposed geological repository at Yucca Mountain as one of the major technical challenges that remain with the program (NWTRB, 2000). The report stated that "DOE has not yet demonstrated a firm technical basis for its present high-temperature 'base case' repository design," and indicated that it looked "forward to the results of DOE work that is under way to evaluate the effects of alternative lower-temperature repository designs on repository and waste package performance" (NWTRB, 2000, p. 2). The SLC's adequacy analysis identified an R&D gap in collecting long-term test data to reduce uncertainty with natural and engineered barrier performance, and indicated that "R&D must continue throughout the active life of the repository to provide data for performance confirmation and to continue to make improvements" in repository and waste package design to reduce uncertainties, increase safety, or reduce life-cycle costs (DOE, 2000g, p. 27).

A panel of the Secretary of Energy's Advisory Board recently completed an evaluation of emerging non-incineration technologies for the treatment and disposal of mixed radioactive wastes. The panel concluded that viable alternatives to incineration exist and should be pursued by DOE, along with basic scientific research to develop a new generation of technologies (DOE, 2000l). DOE's adequacy analysis went even further, identifying alternatives to incineration as "the greatest gap identified among mixed waste technologies" (DOE, 2000g, p. 21).

## Development of Methods and Techniques for Cutting-Edge Characterization and Remediation of Contaminated Media

Several recent NRC studies identified long-term research on the location and characterization of subsurface contaminants, and characterization of the subsurface itself, as high priorities (NRC, 2000c,d). Similarly, the SLC's adequacy analysis found that development of improved sensors and characterization technologies for subsurface contaminants is a significant R&D gap (DOE, 2000g). Long-term R&D to develop improved characterization techniques associated with the deactivation and decommissioning of DOE facilities also was identified as a high-priority need by a recent NRC study (NRC, 2000e); that committee recommended (1) research toward identification and development of real-time, minimally invasive, and field-usable means to locate and quantify difficult contaminants significant to deactivation and decommissioning and (2) research that could lead to development of biotechnological sensors to detect contaminants of interest (NRC, 2000e).

Recent studies also have identified critical long-term R&D needs to develop technologies to remediate contaminated groundwater, soil, and facilities. A number of such studies have concluded that there are significant R&D gaps related to the remediation of subsurface contaminants (NRC, 1999b, 2000c,h; DOE, 2000g). NRC (1999b) concluded with the following summary of the status of DOE's efforts to address its subsurface contamination problems:

> DOE faces the challenge of cleaning up massive quantities of contaminated groundwater and soil with a suite of baseline technologies that are not adequate for the job. Although recent DOE budget projections have indicated that most groundwater at DOE installations will not be cleaned up, federal law requires groundwater cleanup, and political pressure to meet the federal requirements continues. DOE will thus have to continue to invest in developing groundwater and soil remediation technologies. (p. 13)

A recent NRC study on deactivation and decommissioning problems throughout the DOE complex recommended long-term research to develop biotechnological methods to remove or remediate contaminants of interest from surfaces within porous materials; and toward creating intelligent remote systems that can adapt to a variety of tasks and be readily assembled from standardized modules (NRC, 2000e).

## Improvement in Understanding the Movement and Behavior of Contaminants through the Environment

The importance of understanding contaminants' fate and transport in the environment has been duly acknowledged in several recent studies (NRC, 2000c; DOE, 2000g; NRC, 2000i). In a recent report that examined DOE's long-term stewardship responsibilities, a NRC committee came to the following conclusion:

> In some cases, the lack of sufficient pre- or post-remediation characterization of either the wastes or the environments into which they have been placed can render realistic estimation of the effectiveness of contaminant reduction measures nearly impossible. A key question for each site must be "How much characterization is sufficient to overcome this impasse?" A major concern is the adequacy of understanding of the physical and chemical properties of the environment in which contaminants reside and their transport through the environment over time. Mathematical modeling of contaminant fate and transport is an essential tool for long-term institutional management, but its track record to date at DOE sites, particularly where contaminants reside in the unsaturated, or "vadose" zone, has been mixed. This necessitates integration of a science and technology program into both site remediation planning ... and the activities that follow after remediation activities cease.
> (NRC, 2000a, p. 6)

A recent NRC report on the Waste Isolation Pilot Plant (NRC, 2000i) emphasized the importance of establishing accurate baseline information on radioactive materials throughout a geological repository environment so that the movement and behavior of contaminants can be monitored.

R&D on the fundamental approaches and assumptions underlying conceptual modeling of the subsurface also has been identified as a long-term R&D need (NRC, 2000c). The SLC's recent adequacy analysis found that the development of improved understanding of the fate and transport of contaminants in the vadose zone is a significant R&D gap (DOE, 2000g). EM is currently developing a science and technology roadmap for contamination problems in the vadose zone, which should help DOE plan and organize future R&D efforts in this area. Research to improve the understanding of the interactions of important contaminants with materials of interest in deactivation and decommissioning projects was recommended by a recent NRC study (NRC, 2000e).

## Development of Mechanisms for Effective Long-Term Stewardship

The recent report, *Long-Term Institutional Management of U.S. Department of Energy Legacy Waste Sites* (NRC, 2000a), comprehensively examined the capabilities and limitations of the scientific, technical, human, and institutional systems that DOE expects to use under its long-term stewardship program. The authoring committee found that "much regarding DOE's intended reliance on long-term stewardship is at this point problematic" (NRC, 2000a, p. 3) and urged DOE to plan for site disposition and long-term stewardship much more systematically than it has to date. In particular, the committee recommended that "DOE apply five planning principles to the management of residually contaminated sites: (1) plan for uncertainty, (2) plan for fallibility, (3) develop appropriate incentive structures, (4) undertake necessary scientific, technical, and social science research and development, and (5) plan to maximize follow-through on phased, iterative, and adaptive long-term institutional management approaches" (NRC, 2000, p. 4). Among its many recommendations, the committee urged DOE to conduct scientific, technical, and social science R&D to improve its long-term institutional management capabilities. The committee emphasized that long-term R&D should address not only basic technical questions about the behavior of wastes in the diverse environments of the nation's nuclear waste sites, but also the social, institutional, and organizational aspects of long-term management systems. Similarly, a 1998 study from Resources for the Future recommended studies to evaluate institutional alternatives for assuring long-term compliance with institutional controls (Probst and McGovern, 1998).

In a January 2001 report to Congress (DOE, 2001b), DOE identified the following types of technical uncertainties that are important to the success and the assessment of the costs of the long-term stewardship program:

- the nature of the hazards remaining onsite,
- the effectiveness of monitoring,
- the maintenance of barriers and institutional controls,
- the availability of adequate technologies in the future to address residual contaminants,
- the future development of better remedial and surveillance technologies, and
- the long-term management of data.

Long-term R&D on such issues could assist DOE in addressing remaining risks to human health and the environment at closed sites (see also Sidebar 3.1). The recent SLC adequacy analysis also identified a large number of R&D gaps and opportunities in the area of long-term steward-ship (DOE, 2000g, see also Appendix C).

## Determination of the Risks of DOE Wastes and Contaminated Media to Human Health and the Environment

The application of a more risk-based approach to DOE's EQ prob-lems has been a central theme of numerous recent studies (e.g., NRC, 2000f; CRESP, 1999; DOE, 2000g). For example, a recent NRC report on high-level waste (NRC, 1999d) recommended that:

> a risk analysis for the actions recommended above for both HLW calcine and SBW [sodium-bearing waste] should be conducted promptly, and should include a comparison of the risks associ-ated with INEEL HLW calcine and SBW to the risks associated with site inventories of other radioactive wastes. A sufficiently rigorous analysis should be performed to establish the current risks and to assess the changes in risk due to treatment options. (p. xi)

The Peer Review Committee of the Consortium for Risk Evaluation with Stakeholder Participation (CRESP) recently examined the use of risk analysis within EM, and recommended that DOE establish a sound process for developing a risk evaluation methodology that could meet EM's short-term and long-term challenges (CRESP, 1999). CRESP also identified a number of gaps in the knowledge and methods needed to develop such a methodology. SLC's recent adequacy analysis agreed with the needs described in the CRESP report and also identified the need for improved methods for communicating risks to stakeholders as a significant R&D gap in the EQ R&D portfolio (DOE, 2000g).

In summary, what is needed are more accurate, comprehensive, and transparent approaches to assessing and communicating the risks of DOE wastes and contaminated media to human health and the environ-ment so that DOE can make more informed decisions that are accepted by stakeholders. The committee believes that the role for EQ should be to support R&D projects that directly address an EQ problem, such as the relative risks of various treatment options for high-level waste or is-sues associated with the relative risks and public perceptions of dispos-ing of wastes in a geologic repository. This R&D should build upon and leverage other relevant research, such as general research on risk by EPA and other agencies and the Office of Science's research program on health risks from low-dose exposures.

## EXTENDING THE EQ R&D PORTFOLIO BEYOND DOE

One of the questions that DOE asked this committee to address was whether the EQ R&D portfolio should address environmental problems outside DOE that are related to EQ strategic goals. To address this question, the committee undertook a comparative analysis of related R&D efforts outside DOE, as described earlier. The committee answers this question with a qualified "yes." The committee believes it is appropriate for the EQ R&D portfolio to address environmental problems outside DOE, provided that such R&D is directly related to DOE's EQ mission. Earlier in this chapter, the committee concluded that DOE's EQ R&D should be closely coordinated and integrated with relevant parts of DOE's other business lines. Further, it concluded that DOE should leverage the information and technologies developed in programs outside DOE and should make available the information and technologies developed in the EQ R&D portfolio to industry, other federal and state agencies, and other countries.

The committee found no basis to conclude that the EQ R&D portfolio should encompass environmental problems beyond DOE's jurisdiction that are unrelated to DOE's EQ mission. To the contrary, the committee concludes that DOE's current EQ R&D portfolio does not address important long-term EQ problems that are already the responsibility of the EQ business line. There may very well be cases in which spending limited R&D resources on problems outside DOE's EQ mission is appropriate, but deciding when this would be appropriate is less a technical question than a matter of general policy.

**Finding: DOE's current EQ R&D portfolio does not adequately address important long-term problems that are already the responsibility of the EQ business line.**

**Conclusion: It is appropriate for the EQ R&D portfolio to address environmental problems outside DOE if such R&D is directly related to DOE's EQ mission. At this time, however, the EQ R&D portfolio should not address environmental problems beyond DOE's jurisdiction that are unrelated to the EQ mission.**

## MEETING DOE'S LONG-TERM EQ R&D NEEDS

This chapter has discussed the responsibility of DOE for a broad array of R&D activities that can have a dramatic impact not only on DOE's EQ mission but also on its Energy Resources, Science, and National Nuclear Security missions. The EQ responsibilities of DOE are profound, broad, and enduring and they encompass a broad range of issues ranging from the dismantlement of nuclear weapons, with its attendant nuclear materials management, national security, and disposal issues, to the environmental impacts associated with nuclear power and other energy sources. If properly scoped and managed, the EQ R&D portfolio should provide an improved technical foundation for addressing DOE's EQ problems, while setting the highest standards for future environmental stewardship.

The committee pointed out in Chapter 2 that inconsistencies and changes in descriptions of DOE's EQ responsibilities over time may have interfered with developing broad-based support for its EQ R&D efforts. Earlier in this chapter, the committee recommended that DOE establish a long-term, strategic vision for its EQ R&D portfolio. The process of formulating such a vision creates an opportunity to establish clear and consistent objectives that not only provide a baseline for determining an adequate R&D portfolio but that could make it clear that DOE's EQ mission is central to ongoing and future programs throughout the department. One critical goal, in particular, should be to move away from the current "going out of business within the next decade" approach to EQ R&D.

# 4

# ACHIEVING AND MAINTAINING THE LONG-TERM VISION FOR ENVIRONMENTAL QUALITY RESEARCH AND DEVELOPMENT

Chapter 3 described a vision for a different Environmental Quality (EQ) research and development (R&D) portfolio that would have a strong, if not dominant, long-term component. To move towards this vision, the Department of Energy (DOE) will need to redesign and rebalance its EQ R&D portfolio in substantial ways to better focus on its long-term EQ problems.[1] This chapter describes a new portfolio management process that could help achieve these goals.

To be effective, R&D portfolio management must operate within an effective management system, which includes identifying the decision maker (or decision-making group) who will make the hard choices of prioritization, resource allocation, and balance. Portfolio management systems for federal R&D programs also commonly seek out and use input from broadly qualified individuals in generating a comprehensive set of R&D needs and project possibilities. In general, the generation and selection of R&D projects should have inputs from qualified persons both inside and outside the program. This chapter discusses several institutional mechanisms that DOE could use to improve the management of its EQ R&D portfolio, including ways to generate and incorporate such input.

For the most part DOE can implement the recommended new portfolio management process through an evolutionary approach (i.e., by modifying and supplementing existing management processes). The committee believes this is possible because DOE is already using portfolio management techniques (DOE, 2000b,g), and external reviews have found that management based on these techniques is yielding positive results but could be greatly improved (DOE, 2000h). Such an approach avoids disruptive reorganizations and maintains management focus on the goal (i.e., realizing the new R&D vision).

---

[1] As noted previously, the term "EQ problems" refers to the set of technical problems that collectively make up the EQ challenges described in Chapter 2. This is a useful concept in planning an R&D portfolio, because the challenges are very broad, and must be broken down into manageable parts to be addressed by R&D.

## R&D PORTFOLIO MANAGEMENT PROCESS

The primary objective of portfolio management is to ensure that an R&D portfolio is aligned, valuable, and balanced. "Alignment" is intended to ensure that the portfolio supports the strategic objectives and strategic direction of the parent organization (i.e., DOE's EQ mission and objectives). "Value" measures that support in quantifiable terms, such as net social benefit or utility. "Balance" examines whether the portfolio covers the full scope of objectives and approaches or is too narrowly focused on certain categories of R&D, time frames, or topics.

In practice, these three objectives are often treated in sequence. An alignment process typically generates a list of R&D "possibilities" to be considered for funding by a decision maker. The adequacy analysis that identified the extensive list of R&D gaps and opportunities (see Appendix C) was essentially an alignment exercise. A valuation process is one means of prioritizing the list of R&D possibilities so that scarce resources can be applied to deliver the maximum benefit. The Work Package Ranking System (WPRS) that is currently used to select R&D work packages within the Office of Environmental Management (EM) has many similarities to a value-based prioritization system.[2] Balancing a portfolio is a formal process for examining and considering how resources are distributed across critical dimensions and is applied after valuation to offset any imbalances that are inconsistent with overall program objectives. Examples of the types of displays that can be used to evaluate balance include diagrams displaying funding distribution across R&D maturity (Figure 2-6 of DOE, 2000b) and the levels of involvement of universities, national laboratories, contractors, and industries at various stages of R&D maturity (Figure 2-7 of DOE, 2000b). In the following sections, the committee discusses DOE's EQ R&D portfolio management processes in terms of the objectives of alignment, value, and balance.

### Alignment: Generating Improved Project Ideas

Each of the DOE organizations that support EQ R&D has its own process for generating R&D project ideas. These planning processes are designed primarily to gather site and repository needs, which tend to be focused on short-term problems, and to turn these into R&D projects. For example, the participants who determine EM's site needs typically are DOE employees and contractors who are closely involved with the site problems and issues (NRC, 1999a), with some periodic input from the broader

---

[2] DOE does not have a single evaluation method for prioritizing R&D activities across the entire EQ R&D portfolio. Each organization that supports EQ R&D activities has its own process for prioritizing and selecting R&D projects.

technical community, such as from the Environmental Management Advisory Board (EMAB, see Sidebar 4.1). The R&D activities supported by the Office of Civilian Radioactive Waste Management (RW) are identified primarily by DOE staff and contractors at the Yucca Mountain Site, although

---

**SIDEBAR 4.1 Environmental Management Advisory Board**

The Environmental Management Advisory Board (EMAB) was created to provide independent, expert advice, information, and recommendations to the Assistant Secretary of Environmental Management (EM) on issues related to environmental restoration and waste management. Members of EMAB include representatives from state and local governments, tribal nations, environmental groups, labor organizations, private industry, and scientific and technical communities.

EMAB is organized into standing committees, ad hoc committees, and working groups. Three of EMAB's six active standing committees, and one of its ad hoc committees are particularly relevant to this study: The Technology Development and Transfer Committee, the Science Committee, the Long-Term Stewardship Committee, and the Ad Hoc Committee on Technology and Innovation. The missions of these four committees are described below.

The mission of the **Technology Development and Transfer Committee** is to develop implementable recommendations for the Assistant Secretary of EM that can facilitate the development and use of environmental technologies capable of addressing DOE's environmental problems.

The mission of the **Science Committee** is to examine the quality of science in the EM program on behalf of the Assistant Secretary with special emphasis on areas where new science and technology are needed, analyze scientific and technical problems and issues as they arise, and work toward ways to expedite and more efficiently reduce DOE's inherited legacy of environmental cleanup and waste treatment and disposal.

The mission of the **Long-Term Stewardship Committee** is to provide advice and recommendations to the Assistant Secretary on actions EM should take to prepare for and make the transition from its current active programs to long-term stewardship of waste material and property.

The charge for the **Ad Hoc Committee on Science and Innovation** is to examine the linkage between DOE environmental science programs and the long-term stewardship requirements of EM and to recommend how resources and processes could be improved to enable science to be better applied to solving the long-term problems.

Source: Environmental Management Advisory Board web site (http://www.em.doe.gov/emab/)

the Nuclear Waste Technical Review Board also plays a role in identifying science and technology needs for the RW program (see Sidebar 4.2). The Office of Nuclear Energy, Science and Technology (NE) relies on its Nuclear Energy Research Advisory Committee to generate long-term R&D needs (see Sidebar 4.3), although these needs are primarily directed towards nuclear power R&D[3] (and hence DOE's Energy Resources R&D portfolio), because that is NE's overall programmatic focus. In addition, as discussed in Chapter 2, both EM and RW are driven by short-term milestones and deadlines. The short-term drivers and the limited set of participants work together to limit the development of the broad R&D portfolio that was envisioned in Chapter 3.

The recent adequacy analysis of the EQ R&D portfolio conducted by DOE's Strategic Laboratory Council (SLC) was DOE's first attempt to generate R&D project ideas for the entire EQ R&D portfolio (see Appendix

---

**SIDEBAR 4.2 Nuclear Waste Technical Review Board**

Congress created the Nuclear Waste Technical Review Board in 1987 to review DOE's scientific and technical activities pertaining to the management and disposal of the nation's commercial spent nuclear fuel and high-level radioactive waste. These activities include characterizing Yucca Mountain, Nevada, as a potential repository site and packaging and transporting commercial spent nuclear fuel and defense high-level waste.

The board is an independent agency of the U.S. Government whose sole purpose is to provide independent and expert review of the DOE program. The board is composed of 11 members who are experts in science or engineering (including environmental and social sciences) who are selected on the basis of distinguished service. The National Academy of Sciences recommends candidates and the President makes the appointments.

The board has the following primary areas of responsibility:

- makes scientific and technical recommendations to DOE to ensure a technically defensible site-suitability decision and license application;
- advises DOE on the organization and integration of scientific and technical work pertinent to the Yucca Mountain Site; and
- provides an ongoing forum that fosters discussion and understanding among DOE and its contractors of the complex scientific and technical issues facing the program.

Source: NWTRB, 1999.

---

[3] Although the strategic plan included sections on isotopes, space applications, and basic materials research.

---

**SIDEBAR 4.3 Nuclear Energy Research Advisory Committee**

The Nuclear Energy Research Advisory Committee was established in 1998 to provide independent advice to DOE and its Office of Nuclear Energy, Science and Technology (NE) on complex science and technical issues that arise in the planning, management, and implementation of DOE's nuclear energy program. The advisory committee periodically reviews the elements of the NE program and based on these reviews provides advice and recommendations on long-range plans, priorities, and strategies to address the scientific and engineering aspects of the R&D efforts. In addition, the committee provides advice on national policy and scientific aspects of nuclear energy research issues as requested by the Secretary of Energy or the Director of NE. The committee includes representatives from universities, industry, and national laboratories and has the following primary areas of responsibility:

- conducts periodic reviews of elements of the nuclear energy R&D program within NE and makes recommendations based thereon.
- advises on long-range plans, priorities, and strategies to address more effectively the scientific aspects of nuclear energy R&D and stakeholder aspects of the services of NE.
- advises on appropriate levels of funding to develop those plans, priorities, and strategies and to help maintain appropriate balance between elements of the program.
- advises on national policy and scientific aspects of nuclear energy research issues of concern to the DOE as requested by the Secretary or the Director of NE.

Source: Nuclear Energy Research Advisory Committee web site (http://www.ne.doe.gov/nerac/neracoverview1a.html)

---

C). In part because planning has focused primarily on short-term problems, the SLC's adequacy analysis found that the present EQ R&D portfolio does not include a longer-term vision and "strategic elements" and has significant gaps and opportunities (DOE, 2000g). How do these statements fit with the fact that DOE already supports a significant amount of long-term research? The answer lies in the term "strategic," which refers to a plan or method for achieving a goal, including the purposeful allocation of resources. The Office of Science (SC) supports nearly $3 billion in long-term, basic research and scientific user facilities primarily to advance science—not to solve EQ problems (or the problems addressed by DOE's other business lines). Thus, SC research is not "strategically" oriented to EQ purposes, although some of it may provide information useful to the EQ mission. EM also supports problem-oriented, longer-term research in its Environmental Management

Science Program (EMSP). Although all the projects EMSP supports are problem-oriented, they do not, nor were they intended to, comprise a coherent, strategic effort at solving particular EQ problems (see http://emsp.em.doe.gov/). One major reason is that the EMSP budget is small compared to the panoply of scientific problems covered by the program's scope, so that only relatively small, isolated research projects of limited duration are supported.

In summary, the present bias of the EQ R&D portfolio toward short-term R&D (DOE, 2000b,g, h) is to be expected given:

    1.   the way that EM and RW (and to a lesser extent, NE) presently identify R&D needs,

    2.   EM's goal of closing the maximum number of sites (mostly smaller sites) by 2006,

    3.   RW's short-term focus on technical issues associated with site recommendation and licensing,

    4.   the strong emphasis that EM, especially, has placed on getting technologies deployed, and

    5.   declining EQ R&D budgets.

**Finding: The existing processes for generating EQ R&D needs are driven largely by DOE's regulatory mandates, contractor incentives, and short-term goals.**

**Conclusion: The existing R&D planning processes are unlikely to generate the full scope of strategic R&D needed to address DOE's most challenging, long-term EQ problems.**

**Recommendation: DOE should establish a new mechanism within its portfolio management process whose purpose is to develop a more strategic EQ R&D portfolio. The new process should supplement and operate in parallel with existing site-driven processes.**

The primary purpose of the recommended new process, which the committee terms the "Strategic Portfolio Review," would be to identify the gaps and opportunities in the existing portfolio that, when adequately addressed, would encompass the entire spectrum of EQ problems. This Strategic Portfolio Review would be similar to the SLC's adequacy analysis, except that a broader group of experts would participate in the analysis and more explicit criteria that emphasize long-term R&D would be used. Institutionalizing this process is consistent with recent recommendations made by EMAB (DOE, 2000h).

Adequacy would be assessed and gaps and opportunities identified by the judgment of a group of knowledgeable, experienced, and collectively (i.e., as a group) unbiased experts, preferably from both within and outside the DOE community (how to constitute such a group of experts is discussed more fully in the section "Broadening and Deepening the EQ R&D Portfolio"). Gaps and opportunities would be identified using the criteria recommended in Chapter 3 (primarily criteria 1 through 7). The broader scope of the Strategic Portfolio Review would generate a separate, broader, and deeper source of R&D needs on which to solicit, evaluate, and potentially fund additional projects, with emphasis on those that address the highest-priority EQ problems. An example is long-term stewardship, which raises issues involving policies and time scales beyond those now considered in the present R&D portfolio (NRC, 2000a). This requires that the Strategic Portfolio Review focus on the enduring, most challenging problems needing solution, and not on current activities—whether in remediation, waste disposal, or waste management. The expanded set of EQ R&D projects to be considered for funding would consist of projects emerging from the traditional needs processes as well as the new Strategic Portfolio Review.

## Value: Measuring the Magnitude of the Benefit

The gaps and opportunities identified by the Strategic Portfolio Review plus the existing R&D needs processes probably will generate far more demand for R&D activities than can be addressed by current or even greatly expanded resources. Therefore, potential R&D activities will need to be evaluated and prioritized. The goal of valuation is to measure the magnitude of the benefit expected as a result of successful R&D so that scarce resources can be applied to deliver the maximum benefit. Value measures this benefit in quantifiable terms, such as net social benefit or utility. Bjornstadt et al. (2000) have made a strong case for a value-based resource allocation approach for EQ R&D. They suggest risk reduction, cost reduction, and meeting unmet cleanup needs as three components of the potential value of cleanup R&D. They illustrate this approach using a formal non-linear programming model of the Oak Ridge National Laboratory cleanup effort developed for risk analysis (Bjornstadt et al., 1998). Another application of the approach (though used to prioritize relatively short-term R&D needs) is Kaiser-Hill's work at Rocky Flats to identify, prioritize, and mitigate risks to closure project schedule and cost using what was described as an economic optimization approach to decision-making (Kaiser-Hill, 2000). Both of these applications stress the importance of being able to quantify and evaluate risk and uncertainty reduction using "value-of-information" techniques, because the result of R&D is frequently better information as well as new technology. The value-of-information metric for allocating both basic and applied research resources also has been recommended by Fischhoff (2000).

Jenni et al. (1995) discussed an extensive application of a decision support system called the Environmental Restoration Priority System (ERPS). At the heart of ERPS was a multi-attribute utility model, formally elicited from DOE managers, that accounted for six types of benefit:

1.  reduced health risks,
2.  reduced environmental impacts,
3.  reduced adverse socio-economic impacts,
4.  compliance with applicable laws and regulatory requirements,
5.  reduced ultimate cost of clean-up, and
6.  reduced uncertainties relating to risks and costs.

Jenni et al. (1995) used a decision-analytic, value-of-information calculation to quantify the benefits of reducing uncertainty, much as the examples discussed above. Benefits 1-4 had explicit dollar value tradeoffs expressed, such as $200 million to eliminate a 1/10 per year risk of death to the maximally exposed individual, allowing the overall benefit to be translated to equivalent dollars. There were seven full-scale applications of the system between 1988 and 1991, which were "praised in technical review, but strongly criticized by stakeholders external to DOE" (Jenni et al., 1995).[4] A conclusion is that rational value-based systems do work and can in fact deliver most of the promised benefits in use in the DOE EQ environment. It is extremely difficult, however, to convince stakeholders and sites that local interests and site-specific needs can be served by a system explicitly designed with national objectives in mind.

DOE does not have a method for prioritizing and selecting R&D activities across the entire EQ business line, as each DOE organization that supports EQ R&D activities has its own process. The current process used to prioritize EM's R&D needs (OST's WPRS) resembles ERPS as a multi-attribute scoring system. RW uses a "focused approach" that funds the R&D work required to allow submittal of the site recommendation report and, if the site is selected, the license application to the Nuclear Regulatory Commission. NE considers potential life-cycle cost savings, potential reduction in environmental safety, and health risks, technical viability, and regulatory requirements to prioritize its R&D investments.

Because EM supports over 80 percent of the R&D activities within the EQ R&D portfolio, and because the WPRS is a fairly well documented, formalized process, the committee discusses it at some length in the following paragraphs. The purpose of this discussion is to explain why the WPRS currently emphasizes short-term R&D needs, and to suggest ways that the WPRS could be modified to be useful in identifying long-term R&D needs. The ranking system is based on five criteria:

---

[4] It should be recognized that the sites also have reasons for parochialism, as discussed in Chapter 2.

1.   project baseline summary "value" (i.e., a measure of the total life-cycle costs of the baseline technologies to be replaced by a given work package, which is intended to reward those work packages that address high-cost projects and that can be employed at more than one location);
2.   future technology deployments (i.e., the number of times the technologies within a work package are expected to be deployed);
3.   response to site science and technology needs (i.e., the number of site-identified priority needs addressed by the work package);
4.   addressing technical risk (i.e., a measure of the baseline technology's technical risk); and
5.   technology cost savings (i.e., a measure of the potential ability of the work package to achieve cost savings compared to baseline technologies).

The WPRS offers a number of major benefits relative to earlier methods used in EM, including being based on end-user life-cycle planning data, better understanding of work package benefits, and direct alignment with EM's four corporate performance measures.[5] The ranking system has been favorably reviewed by EMAB (DOE, 1999b) and appears to do a good job of concentrating on site needs and deployable technologies.

There are, however, several features of WPRS that limit its usefulness as a valuation tool for the types of R&D that are under-represented in the EQ portfolio (DOE, 2000g). Because EM's four corporate performance measures are understandably oriented toward near-term accomplishments, the ranking system inherits that near-term focus. In particular it is directly tied to needs articulated by the sites, who by their nature have a more operational, shorter-term focus; one would not expect them to focus on needs beyond 2006. Also, the primary incentive for most sites is to meet their legal and contractual obligations, so a new technology that offers significant cost reduction but might delay the program is typically unwelcome.

There are additional reasons why the WPRS is currently not well suited for evaluating the R&D oriented at strategic R&D. The five criteria included in the ranking system are not as reflective of society's priorities as they are of EM management's performance measures. While this is by design, the six criteria used in the ERPS, for example, are a better reflection of national needs. Other better alternatives would be to use the seven EQ objectives listed in the R&D Portfolio Overview (DOE, 2000i) or the five refined EQ objectives used in the adequacy analysis (DOE, 2000g), including stewardship. A guiding principle for the design of an improved evaluation system should be that it could apply equally well to all areas of the EQ portfolio, not just to EM.

---

[5] (1) number of new technology deployments, (2) life-cycle cost reduction from use of science and technology, (3) number of high-priority needs that are met, and (4) reduction in critical pathway milestones and waste stream technical risk (DOE, 2000n).

In addition, there appears to be some redundancy (or lack of independence) among the five WPRS criteria. For example, "site technology needs addressed" also is reflected in number of deployments, technical risks addressed, and technology cost savings. Having it as an additional criterion could lead to double counting. In addition, working on "high value" project baseline summary elements is not an end in itself, unless it leads to large improvements in the cost savings or technical risk categories. As an example, a diffuse work package that addressed many maximum-value project baseline summary elements but had trivial impact on technical risk or cost savings could have the same score as a tightly focused project that had maximum impact on reducing technical risk in one very low cost (but critical) project baseline summary element, although only the latter would produce any benefit. Although projects with no real potential benefits are unlikely to be proposed, it would be better for the scoring system to rate them very low. Finally, "high-value" project baseline summary elements are actually based on cost, not value, so this criteria is questionable as defined. For these reasons, the use of these five criteria in an additive multi-attribute utility model is open to question.

It is unclear how the probability of technical success of the R&D projects in the work packages enters into the WPRS evaluation method. Typical R&D evaluation methods (NRC, 1999a) involve benefit, cost, and probability of technical success in the prioritization process. It could be that this is intended to be captured by a number of deployments, but that is not the same. It is also unclear how the value of time (e.g., discounting of future cost and benefits) is dealt with in the WPRS process. Finally, the value of information, which was a key feature of several of the other systems cited, is not evident in the WPRS.

In summary, the WPRS as currently implemented is EM-specific and is not well suited to evaluating long-term R&D. Something more like the examples of Bjornstad et al. (2000) and Jenni et al. (1995) is needed for more strategic R&D. Although separate evaluation methods may be needed in the short term, it may be possible for the present WPRS to evolve to one that addresses the concerns outlined above, that works for all areas of the EQ portfolio, and that is equally robust for both long- and short-term R&D, using many of the same data but with a modified algorithm along the lines of those discussed above. It would be highly desirable for the probability of technical success of work packages (or projects within work packages) to be made explicit, the treatment of time preference clarified, and the value of reducing uncertainty captured by the improved system. It also might be valuable for EQ to learn how industry treats probability of technical success, especially industries where technical risk is high (e.g., pharmaceuticals). Finally, as discussed in Jenni et al. (1995), to be effective such a methodology must be designed and applied openly and objectively so that all stakeholders can provide comments and understand what is being done.

<u>Finding</u>: The current Work Package Ranking System is heavily biased toward activities that are site generated and connected to the present baseline plans. Moreover, it is by design EM specific and therefore does not apply to other parts of the EQ R&D portfolio.

<u>Conclusion</u>: The current Work Package Ranking System is unlikely to be effective in prioritizing R&D activities designed to address the long-term strategic gaps and opportunities identified in the Strategic Portfolio Review discussed above, especially those not within EM.

<u>Recommendation</u>: DOE should develop and implement an evaluation method to address more strategic R&D for the entire EQ R&D portfolio. In the short term, it could be entirely separate from the EM's Work Package Ranking System, but in the longer term a new approach is needed that works for both site-driven and strategy-driven activities and is applied within all areas (i.e., EM, RW, NE) of the EQ R&D portfolio.

Several good models for such a system that have been applied to elements of the DOE EQ portfolio but that are not EM specific have been discussed above.

### Balance: Ensuring Adequate Attention to Diverse Objectives

A common experience in life is that the urgent overwhelms the important. It is typical in business R&D organizations that, without strategic guidance, requests for short-term product and process improvements can exhaust the available resources. The results of SLC's adequacy analysis indicate that this is likely true for EQ R&D. Balancing can offset such forces by examining how resources are distributed across various critical dimensions, such as how R&D is distributed across the strategic objectives of the parent organization, across time frames, across risk versus return, or across R&D stages.[6] Balance is also important when the potential value of one objective is so large that projects addressing it tend to dominate projects addressing the other objectives, as might be the case for efforts to reduce the cost of some of DOE's most expensive cleanup problems. The diversity of these considerations demands that DOE seek and use the breadth of advice described in the next section.

One of the most common balance metrics in business is the relation of technical risk (or the probability of technical success) to return (i.e., value). Allowing DOE to track technical risk and value would be another benefit of

---

[6] For the EM portion of the EQ portfolio, how R&D is distributed across focus areas and sites also may be useful to DOE decision makers.

making probability of technical success an explicit part of the evaluation process.

Three of the 10 adequacy criteria (8-10) developed in Chapter 3 pertain to elements of portfolio balance. The most fundamental balance issue is the proportion of the budget that should be allocated to strategic R&D as opposed to R&D driven by short-term needs. There is no simple answer to this question. For example, the appropriate proportion of strategic R&D would be quite different in DOE's Energy Resources R&D portfolio, where nearly all commercialization and deployment is done in the private sector, than in the EQ R&D portfolio, where deployment and application are mostly internal to DOE and its contractors. A number of recent analyses have concluded, however, that more strategic R&D is needed to adequately address DOE's EQ objectives (DOE, 2000g,h). In addition, the SLC's adequacy analysis (DOE, 2000g) examined the funding distribution across the technology maturity spectrum and concluded that it is unbalanced. The committee discusses methods for evaluating EQ R&D funding balance in Chapter 5.

## INSTITUTIONAL MECHANISMS

This section discusses implementation of the recommendations made above and offers additional recommendations related to institutional mechanisms that could be used to make the EQ R&D portfolio more effective in addressing long-term problems, including the personnel needed to carry out the Strategic Portfolio Review and a new approach to long-term EQ R&D that could be added to existing programs.

### Broadening and Deepening the EQ R&D Portfolio

Several reviews of the EQ R&D portfolio have concluded that the portfolio is too narrowly focused on short-term problems and needs a broader perspective to address the most challenging EQ problems and to limit contamination and materials management problems in ongoing and future DOE operations. For the portfolio to adequately address DOE's most challenging EQ problems, the agency must gather input for the Strategic Portfolio Review from a much wider range of people than it customarily involves in its program management (see for example: DOE, 2000h; NRC, 1998, 1999a). To achieve this, several kinds of individuals are needed. Although individual experts are almost by definition narrow in scope, a well-chosen group of informed individuals working together can achieve a very broad perspective. The following categories illustrate by example the range of knowledge individuals can bring so that collectively they match the breadth of the EQ problems to be addressed:

- practical, problem-oriented, and technically trained experts with relevant experience (e.g., from a relevant industry or a foreign, federal, or state agency with a similar mission);
- applied researchers in relevant technologies (e.g., radiation hardening of sensors);
- applied researchers in generically important technologies (e.g., robotics);
- "basic" researchers in relevant areas (e.g., actinide chemistry);
- "basic" researchers in generically important sciences (e.g., physical chemistry);
- individuals possessed with a broad, long-term perspective of where technology, science, and/or environmental problems and policies are trending but who are from outside the DOE family of employees, national laboratories, contractors, and others whose interests might appear to represent substantial conflicts of interest; and
- technically qualified individuals representing nongovernmental organizations and other stakeholders.

Participation must be broadened carefully to ensure the success of the strategic review process. That is, the composition of the group should balance the need for a diversity of expertise with the need for an efficient process. The intent is to select a group of individuals who collectively are not predisposed in favor of the existing portfolio or any particular approach to solving a specific EQ problem (that is, individual biases in the group will be balanced). Consideration should be given to experts from other countries who have the needed expertise because these individuals can bring valuable perspectives to the review and are less likely to have a stake in the outcome. Qualified individuals from outside the program bring a broader perspective and often can look "outside the box" for new approaches. Of course strong input also is needed from DOE staff who have responsibility for accomplishing the EQ mission and special familiarity with the difficulties they face, and who must have the final word if they are to be accountable.

The purpose of the strategic review would be to attain the broader input and perspectives that seem lacking (DOE, 2000h). The group should be able to identify the full scope of R&D needed to solve EQ problems and to give rough priority rankings. DOE might chose to experiment with organizational approaches to find the best way to gather and use the information such a group can provide. For example, the group could be established as a new, ongoing EQ R&D advisory board. Another possible approach would be create this group largely from members of EMAB, the Nuclear Waste Technical Review Board, and the Nuclear Energy Research Advisory Committee (see Sidebars 4.1, 4.2, 4.3). This group would differ from EMAB, the Nuclear Waste Technical Review Board, and the Nuclear Energy Research Advisory Committee in that it would focus on the EQ R&D portfolio, have continuity to see how its recommendations were carried out, and be much more integrated into and a part of the regular EQ R&D management process. Such a board would report to the manager of the EQ R&D portfolio, and if such a manager does not exist, to the line managers of its components (e.g., the Assistant Secretary for EM, the Director of RW, the Director of NE). It would meet at least annually as needed to develop a strong R&D portfolio.

**Conclusion**: **An independent advisory group representing a broad spectrum of expertise and experience is necessary to assure a sustained, high-quality EQ R&D portfolio.**

**Recommendation**: **DOE should establish an independent planning and review board specifically focused on the EQ R&D portfolio, with membership composed of leaders in the scientific and technical community, including experts from industry, academia, national laboratories, and affected communities. The purpose of this board would be to recommend to DOE management and justify in terms of program and mission a world-class R&D portfolio with the breadth and depth to address EQ problems.**

### Technical Qualifications of Staff

The EQ R&D portfolio represents a highly technical activity, and must be managed by a staff with strong technical qualifications (DOE, 2001a). The portfolio management techniques and the independent advisory board recommended above do not reduce the need for strong in-house technical management, because DOE staff still must make the final decisions. It takes considerable technical insight to identify a practical problem in the field and then determine whether current technology can resolve it, and if not, to translate the problem into researchable questions and eventually into R&D projects leading to understandings or technologies to mitigate the problem. EQ R&D managers must make researchers aware of how their work could lead to solutions to critical problems and convince them to pursue such useful results. Conversely, similar insight is needed when examining a proposal for fundamental research to visualize its application to practical problems—and to know whether and how it should be funded. Finally, EQ R&D managers must work with operators in the field to implement R&D results. Thus, EQ R&D managers need the technical depth and breadth to span the conceptual range from the field to the lab and back. A DOE staff with such depth and breadth would be able to take advantage of technical advice from outside groups as recommended above.

Individuals with such talents are rare, but two approaches can deal with this problem. First, management can partially compensate for the unavoidable limitations of individuals by bringing a broad range of views into management processes. Second, because universities generally do not train individuals with such a comprehensive grasp of problems and solutions, DOE might have to adopt institutional approaches to develop them. A recent EMAB report on the role and status of basic science in EM recommended that EM address this issue by developing "operational procedures" for OST staff positions similar to those used by EPA and that OST establish requirements for those positions that reflect their scientific and technical nature (DOE, 2001a). The new approach to addressing DOE's long-term EQ problems described below also could help develop such people.

## An Approach to Addressing DOE's Most Challenging, Long-Term EQ Problems

Chapter 3 described a set of criteria that could be used to evaluate the adequacy of DOE's EQ R&D portfolio. These same criteria can be used to help design a new approach to EQ R&D that could improve its effectiveness in addressing these long-term currently intractable problems, and which would supplement existing R&D programs. For example, the approach should:

- Address critical R&D gaps needed to address EQ goals (and when appropriate, to support the accomplishment of related DOE and national missions).
- Encourage the development of alternatives to technologies that are costly, inefficient, or pose high technical risk.
- Produce results that could transform the understanding, need, and abilities to address currently intractable problems, thus enabling breakthrough technologies.
- Lead to improved performance, reduced human health or environmental risks, decreased cost, and advanced schedules.
- Help leverage other R&D, such as the Environmental Management Science Program.
- Help to narrow and bridge the gap between R&D and application.
- Improve the balance of long- versus short-term research.
- Involve a diversity of participants from academia, national laboratories, other federal agencies, and the private sector, including students, postdoctoral associates, and other early-career researchers.
- Include a balance of annual new starts, extensions of promising R&D, and periodic new initiatives.

The committee believes that in order to meet these criteria, a significant fraction of R&D should be conducted in organizationally separate units to help maintain a focus on long-term results. Each of these units would be strongly coupled to an important, currently intractable EQ problem and evaluated according to progress on solving the problem, but not strongly coupled to short-term program needs. Based on these general criteria and considerations, the committee arrived at the following finding and recommendation. The committee then describes some of the characteristics of the recommended approach.

**Finding**: **Given the long-term nature of many of DOE's EQ problems, there is a need to develop sustained support for R&D activities to solve such problems.**

**Recommendation**: **DOE should implement a new approach to provide longer-term funding for organizationally separate, integrated, and coordinated R&D activities (i.e., R&D centers) designed to solve well-defined, high-priority EQ problems.**

The most important element of the recommended approach is that each R&D center[7] should focus on providing longer-term support for solving a particular long-term EQ problem, specifically countering the "going out of business within the next decade" philosophy that has permeated some views of the EQ portfolio (see discussion in Chapter 2). Here it is appropriate to differentiate and clarify what is meant by the phrase "a particular long-term EQ problem" with respect to other concepts, such as "EQ challenge" and "focus area." The committee's term "EQ challenge" (see Chapter 2) refers to the broad challenges facing DOE in its EQ mission area. The management of EM has organized its R&D effort to address some of these challenges into "focus areas" and "crosscutting programs." All of these are very different from what the committee means by a "problem." First, they are much broader and more general. Second, they do not refer to an integrated, coherent effort to solve a problem, but to collections of R&D efforts. Third, focus areas and crosscutting programs sometimes mean problems, but usually mean R&D activities—an unfortunate confusion. The universe of problems that might be assigned to R&D centers, taken together, overlaps with the long-term component of all the EQ challenges and of all the focus areas and crosscutting programs. What "problem" means here is an issue, a hindrance to progress, that is appropriate to be

---

[7] The committee refers to the organizations carrying out the integrated and coordinated R&D efforts as R&D centers to indicate that the whole of each is greater than the sum of its parts, i.e., that integration and coordination to focus on a central, often multidisciplinary problem leads to synergy and a holistic solution. This synergy could be achieved through a variety of organizational approaches. For example, the centers could have a virtual aspect, using technology to involve experts at various locations. Thus one sort of balance to be struck is that between the benefits of daily face-to-face collaboration and achievement of critical mass in that sense versus the achievement of a different sort of critical mass by involving many geographically dispersed experts. A center could not be completely virtual, however. For example, there would have to be a locus of coherence, accountability, and problem ownership.

addressed by a single integrated, coordinated, focused R&D effort of the scale that can be supported realistically.

DOE could initiate the new approach by identifying a few well-defined high-priority problems and releasing a set of competitive requests for proposals calling for integrated and coordinated R&D activities to solve each problem. As discussed above, each problem would be a manageable part of the larger EQ challenges of Chapter 2 (i.e., it would represent a barrier to program progress), not just a scientific question. The problems to be addressed could be based on (i.e., perhaps a subset of) the gaps and opportunities identified by the Strategic Portfolio Review. At its core each problem would have at least one unanswered scientific (including social science) or technical question (i.e., this is why R&D is needed) and the centers would pursue these questions in their problem context and in consultation with users, not as pure technical questions. Assigning an R&D center a real-world problem would give it flexibility to choose among technical approaches, indeed to choose more than one if appropriate. Success would be measured in terms of progress in solving the problem. The problems should not be too global (e.g., "reduce the amount of radioactive waste") or too narrow (e.g., "make a particular technology work"). Sidebar 4.4 describes an example of a possible type of R&D center based on a recommendation from a recent NRC report (NRC, 2000c).

The problem would determine the disciplines and the types of R&D (e.g., fundamental research, applied research, and development) and the number of investigators needed in each R&D center. Most centers probably would be highly multidisciplinary and would involve different types of R&D. For example, a center might include fundamental research, applied research, and perhaps some engineering research to demonstrate the efficacy and practicality of an idea. In addition, centers would be encouraged to involve participants from other agencies and other countries where appropriate. For some problems (e.g., those with high technical risks or of particular importance to EQ mission success), DOE might consider funding more than one center in order to increase the likelihood of success.

The R&D center would be expected to frequently consult with and involve its user-clients, which would generate a "technology pull" from them. In a sense, the center thus would become a co-owner of the problem. Large downstream development funding might be needed to achieve application in the field, but the center would take responsibility for seeing its own results applied. The R&D center would thus support a technology's maturation through the development process (e.g., by consulting on problems that arise and perhaps doing some supportive research). In other words, the center would help bridge and narrow the gap between R&D and application (NRC, 2000b). R&D centers also would be encouraged to involve students and postdoctoral fellows to achieve the educational and training function described in Chapter 3 and mentioned above. Finally, the centers would be encouraged to coordinate and cooperate with related R&D activities,

---

**SIDEBAR 4.4 An Example of an R&D Center to Solve Subsurface Contamination Problems**

A recent NRC report, *Research Needs in Subsurface Science, U.S. Department of Energy's Environmental Management Science Program*, included a recommendation very similar to the R&D centers recommended by this committee. Below is a quote from the Executive Summary of that report, describing field sites that could be used to address subsurface contaminants problems:

*The committee recommends that [Environmental Management Science Program] program managers examine the feasibility of developing field research sites as one program component. Such sites could attract new researchers to the program, encourage both formal and informal multidisciplinary collaborations among the researchers, and facilitate the transfer of research results into application. These field sites could include contaminated or uncontaminated areas at major DOE sites; analog uncontaminated sites that have subsurface characteristics similar to those at contaminated sites; and even virtual sites comprised of data on historical and contemporary contamination problems. These sites could be established by the program itself or in cooperation with other research programs.*

*The establishment of field research sites is potentially expensive, especially if the sites are located in contaminated areas. Consequently, the establishment of such sites will require additional budget support beyond that required to fund individual research projects, and well beyond the amount of funding available to the program for new starts in fiscal year 1999. Moreover, the use of such sites will have to be evaluated periodically to determine whether they are adding value to the research effort, particularly given the cost of such sites relative to the total size of the program budget. (NRC, 2000c, p. 8-9)*

Clearly, this recommendation is consistent with the committee's recommendation for R&D centers. A field research site could be the focus for a center that addressed a relevant set of problems. Because such a field-based center would receive R&D funding directly, it would differ from many field research sites established as user facilities. It also could provide a way to generate synergy with related Environmental Management Science projects and serve as a focus for center participants. It should be noted that the best R&D organizations are not always situated at the sites most in need of the R&D results. Centers should be awarded, established, and managed such that the best organizational talent is brought to bear on the problem.

---

including EMSP and SC projects, work in other agencies, and work in other countries.

Although the R&D centers would deal with currently intractable problems they would nevertheless be evaluated in terms of problem solution. The centers would be strongly encouraged to seek breakthroughs, even at the cost of some technical risk. To mitigate such risk and improve the probability of overall program success they would be responsible for seeking and developing alternative parallel paths. Each center would be overseen by an independent technical advisory committee familiar with the problem being addressed. Each center would be evaluated regularly on the basis of its progress in solving the assigned problem and overall technical soundness of its R&D. For funding to continue, the center would have to demonstrate first, that it is making progress in solving problems and, second, that it is sound scientifically and technically. For credibility, centers not making adequate progress toward solving EQ problems should be terminated by DOE, not by the Office of Management and Budget or Congress. Those R&D centers making adequate progress could be renewed if the problem remained important.

The committee did not examine in detail the funding that might be required for the R&D centers, but based on its members' experience as R&D managers and knowledge of DOE's EQ R&D portfolio, it believes that an appropriate figure for each center would be approximately $1-4 million annually for five years. The suggested funding range is meant to balance at least two considerations: (1) given the limited funds available and the desire to start several such centers, each must be small; and (2) on the other hand, each R&D center should be large enough to make progress toward problem solution. A problem calling for multidisciplinary R&D might need a larger R&D center. Such considerations also should help identify the problems to be addressed (i.e., the problems must be of a size appropriate to available funding).

Because the approach to EQ R&D recommended here would be new to DOE it should start small and grow only as long as justified by the problems. With the first set of R&D centers well underway, DOE could take steps to enlarge the program by selecting another small set of high-priority problems and repeating the process, fine-tuning the new centers to take advantage of lessons learned. The portfolio of problems addressed would grow as long as problem owners, stakeholders, DOE management, and other decision makers supported such efforts. By this process of continual improvement, EQ could build a portfolio of expertise to apply to its most important problems.

## Need for Coordination of EQ R&D

R&D portfolio management, a recent innovation at DOE, begun in 1999 by its Under Secretary, covers the department's four programmatic business lines, with each having an R&D portfolio. The goal is "to integrate and strengthen the planning, management and administration" of the $8.0 billion DOE R&D enterprise (DOE, 2000i, p. 1). The means for achieving this goal are unclear, however. DOE documents do not address how the goal will be achieved with any specificity or depth (DOE, 2000b,i). DOE is aware of this deficiency; both the SLC review of the portfolio (DOE, 2000g) and a recent letter report by EMAB (DOE, 2000h) found that it was a good start but needed to be improved to achieve its goal.

As presented by DOE, the portfolio concept itself raises questions about whether it can achieve its stated goal. First, the portfolio is presented as being only a "context" for R&D: the portfolios "have no funding per se; they provide the context within which the funded programs and offices manage and execute their funding" (DOE, 2000i, p. 4). The R&D portfolios are descriptive tools, and no decision, budget, or priority-setting authority is associated with them. Because of this, no accountability is associated with them. These limitations are common to all the R&D portfolios, including the EQ portfolio.

EM, and its Office of Science and Technology, manages the great majority of the EQ R&D portfolio. Consequently, line managers could coordinate this part of the portfolio if given incentives to do so. However, other offices (see Sidebar 1.4) conduct some of the portfolio. In addition, as discussed below, there is much research in the Science portfolio in disciplinary areas of great interest to EQ programs, such as work on the movement of groundwater and on bioremediation.

DOE has taken a first, important step toward integrating its R&D programs through portfolio analysis. However, DOE's portfolio concept (i.e., as a context only) offers no way to reach across organizational or portfolio lines to coordinate R&D. The portfolios do little to cross DOE's existing organizational stovepipes. The relationship between EM and SC (which wholly owns DOE's Science portfolio) illustrates the situation. DOE's 2000 strategic plan directs SC to "advance basic research and the instruments of science that are the foundations for DOE's applied missions... to support long-term environmental cleanup and management at DOE sites...." (DOE, 2000f, p.7). In other words, although it supports applied missions, its research is "basic" (i.e., it looks within science for its research questions and justifications), whereas EQ R&D must address external problems directly. Put another way, SC sees research as an end in itself, but for EQ research is a means to an end. As discussed briefly in Chapter 3, these different world views make cooperation and coordination correspondingly difficult, and unlikely without conscious, continual effort. The Environmental Management Science Program, which is administered jointly by EM and SC, demonstrates that such cooperation and coordination are possible, however.

DOE's portfolio approach also cannot compare programs between portfolios. For example, there is no common system for setting priorities or evaluating results. As the preceding example shows, different portfolios have very different metrics for success and different definitions of what is a worthwhile problem. Portfolio coordination and management are needed at high and low levels. At higher, strategic levels they are needed to deploy resources on the main problems. At lower, tactical levels they are needed to minimize duplication and overlap, to create synergies, and to ensure stakeholder involvement. All this is another way of saying that there is a need for alignment, value, and balance across and within portfolios, and at both strategic and tactical levels (realizing that because one cannot balance within the smallest program elements, balance is sought among such elements).

The generation of the R&D portfolios is a sound accomplishment that might provide a starting point for coordination. However, as presently described, DOE's portfolio management approach seems unlikely to achieve its goals. To be effective the portfolios will need to become a management tool, not just a descriptive tool. That is, the portfolio process would need to include explicit management functions and capabilities, especially accountability.

**Finding: There is little evidence of effective coordination within the EQ R&D portfolio (e.g., for communication of results or for recommendations on priorities). Furthermore, there is little evidence of effective coordination between R&D portfolios.**

**Conclusion: At present DOE's R&D portfolio process is unlikely to achieve its goal to integrate and strengthen the planning, management, and administration of its $8 billion R&D enterprise.**

DOE recognizes that "the portfolio process would benefit from improved coordination and a more integrated approach to ... interportfolio activities" (DOE, 2000b, p. xiii). Although an understatement, this does indicate that the process may improve. Accordingly, the committee is reluctant to make a specific organizational recommendation that might limit DOE's options. In the past, the Under Secretary chaired a group, the R&D Council, whose members included the DOE leadership responsible for each of the four R&D portfolios. Although the status of the R&D Council remains uncertain following the 2000 election, the charter of such a group would allow it to oversee coordination of the EQ portfolio, as well as coordination between portfolios. Because its members have other duties and loyalties, however, such a group alone is an unlikely vehicle for coordination. The committee believes that such a group could serve as a forum for discussion and agreement on plans for coordination developed by the Under Secretary's staff.

The larger issue goes beyond specifics to whether DOE intends the portfolios to be more than a context (i.e., whether they should be actively managed). If they are to remain only descriptive (i.e., to reveal problems but not to address them), some other means for achieving the goals of improved R&D management must be found.

# 5

# THE LEVEL OF INVESTMENT IN DEPARTMENT OF ENERGY ENVIRONMENTAL QUALITY RESEARCH AND DEVELOPMENT

In previous chapters the committee has identified the need for focused, vigorous, and sustained, research and development (R&D) activities to address the long-term problems faced by the Department of Energy's (DOE's) Environmental Quality (EQ) business line. The committee also has developed criteria to evaluate the adequacy of the EQ R&D portfolio, described the principal elements of the portfolio, and described how DOE could achieve and maintain a more effective, long-term R&D portfolio. This chapter identifies and refines measures that could be useful in determining an appropriate level of R&D investment.

EQ is DOE's second most expensive business line, accounting for $6.7 billion of the $19.7 billion DOE budget for fiscal year 2001 (see Table 5.1). The annual investment in EQ R&D is the smallest of the four business lines, however. For fiscal year 2001, funding for EQ R&D was about $298 million (4 percent of DOE's EQ budget), versus about $1.3 billion for R&D on energy resources (52 percent of DOE's Energy Resources budget), $3.4 billion for R&D on national security (49 percent of DOE's National Nuclear Security budget), and $3.0 billion for research on "science" (nearly all of DOE's Science budget). The smaller R&D investment in the EQ business line relative to that in the Energy Resources and National Nuclear Security business lines suggests that decision makers in DOE and Congress do not view EQ R&D as a high priority. Another indication that EQ R&D is not a high priority is a comparison of the trend of DOE EQ R&D spending with the trends in R&D spending in DOE's other business lines (see Figure 5.1). Figure 5.1 shows that from FY 1999 to FY 2001, R&D spending increased significantly in every DOE business line except EQ, where it was reduced by more than 8 percent over this same time period. The small increase in fiscal year 2001 suggests that the declining EQ R&D funding trend that occurred from 1995 to 2000, during which funding was reduced by nearly 50 percent (DOE, 2000b), may have stabilized or even reversed slightly. Even so, Figure 5.1 shows that in fiscal year 2001 R&D funding by DOE's other business

TABLE 5.1 R&D Funding of Selected Federal Agencies (fiscal year 2001, current appropriations)

| Agency/Business Line | Total Budget ($ billion) | R&D Budget | R&D as % of Budget |
|---|---|---|---|
| Environmental Protection Agency | 7.8[a] | $686 million[a] | 9 |
| Department of Defense | 288[a] | $41.8 billion[a] | 15 |
| Department of Energy (total) | 19.7[b] | $8.0 billion[a] | 41 |
| Science | 3.2[b] | $3.0 billion[a] | >90 |
| National Nuclear Security | 7.0[b] | $3.4 billion[a] | 49 |
| Energy Resources | 2.5[b] | $1.3 billion[a] | 52 |
| **Environmental Quality** | **6.7[b]** | **$298 million[c]** | **4** |
| Corporate Management, Other | 0.3[b] | - | - |

Data Sources:
[a] AAAS, 2001.
[b] Department of Energy Office of Chief Financial Officer (http://www.cfo.doe.gov/budget/02budget/3-pager.pdf)
[c] K. Chang, DOE (personal communication).

lines increased significantly more than in EQ. These data are an indication that decision makers in DOE, the Office of Management and Budget, and Congress may not fully understand the magnitude and duration of many of the challenges faced by the EQ business line and the potential value of long-term R&D to address such challenges. The small amount of EQ R&D investment also is consistent with a business line that is required to meet many important short-term milestones, including regulatory requirements and the goals of DOE's accelerated clean-up plan (DOE, 1998). The data also may reflect the perception that past R&D investments have not resulted in many deployments of new technologies (U.S. House of Representatives, 2000).

Determining an appropriate level of R&D investment requires clarity on three principles:

• **The level of investment depends on the scope of DOE's EQ mission.**

• **The level of investment must take into account the balance to be drawn between spending limited resources on R&D and other possible applications of those resources in meeting other EQ goals and commitments.**

- **There are no formulas or mechanistic ways that by them-selves provide or justify a specific funding level recommendation.** Nonetheless, the committee discusses two techniques (benchmarking and investment indicators) that DOE should use as guides in determining an appropriate level of EQ R&D investment.

Each of these principles is discussed below.

### DEFINING THE GOALS AND OBJECTIVES OF THE EQ MISSION

Broad based support for EQ R&D investment first requires a clear and compelling presentation of, and commitment to, the goals and ob-jectives of the EQ mission. As discussed in Chapter 2, there is currently some lack of clarity and consistency in DOE's EQ goal and objectives, and this deserves careful consideration and clarification. The similar and

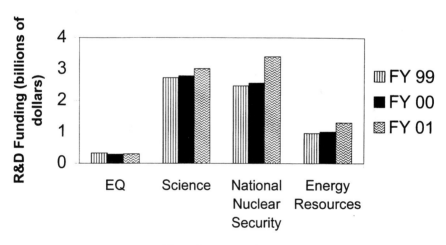

**Business Line**

FIGURE 5.1. DOE Business Line R&D Spending by Year (in billions of dollars). Data for fiscal year 2001 (FY 01) are estimates based on appropriations. Data from 1999 and 2000 for all business lines are from DOE's R&D Portfolio Over-view (DOE, 2000i). Data from 2001 for the National Nuclear Security, Science, and Energy Resources business lines are from AAAS (2001); 2001 data for EQ are from K. Chang, DOE (personal communication).

overlapping but clearly differing descriptions of the EQ goal and objectives over the past three years makes it difficult to both identify and defend the appropriate levels of R&D investment.

To resolve the matter, the committee has recommended that DOE develop strategic goals and objectives for its EQ business line that incorporate a more comprehensive, long-term view of DOE's EQ responsibilities. If DOE accepts this recommendation, it will almost certainly need to re-examine the level of EQ R&D investment. Once clear and enduring goals and objectives are defined, methods such as those discussed later in this chapter should be a base for analyzing whether the level of R&D funding is appropriate for meeting DOE's EQ responsibilities. Of course, funding alone cannot ensure that EQ R&D will be effective. Effective portfolio management, such as that described in Chapter 4, is an important first step in ensuring that DOE's EQ R&D investments are used effectively.

## BALANCING R&D INVESTMENTS WITH OTHER IMPORTANT EQ NEEDS

After clear and appropriate goals and objectives have been defined, determining the level of R&D investment will require difficult tradeoffs from DOE managers and others. There are many important short-term issues that call for high-priority allocation of funds. In EM, often reinforcing or driving these needs are milestones associated with existing compliance agreements between DOE and state environmental regulatory authorities, and concomitant expectations of the affected communities and their representatives. Congressionally mandated milestones are major drivers of much of the Office of Civilian Radioactive Waste Management's (RW's) program activities. In such situations, allocating funds to R&D can be seen as taking resources away from meeting short-term requirements or compliance agreements to support activities that are, by their very nature, longer term and more uncertain in their ultimate benefits. It is incumbent upon DOE leadership to make clear to all EQ stakeholders the value of a robust and sustained R&D portfolio in addressing the most challenging EQ problems.

DOE has made initial steps in recognizing this required balance through the creation of the Environmental Management Science Program (EMSP), the focus areas and crosscutting programs within EM's Office of Science and Technology (OST), and R&D programs in RW and the Office of Nuclear Energy, Science and Technology (NE). However, most of these EQ R&D programs have not been characterized by a history of "strong, stable funding for a portfolio of research investments that is diverse in terms of funders, performers, time horizons, and motivations" that is needed for effective "capitalization" of R&D results (COSEPUP, 1999a, p. 4).

For example, decreases in funding for the EMSP program, together with significant "mortgages" imposed on the program from previous years' awards that were not fully funded (see discussion in NRC, 1997), have significantly reduced the number of new grants that can be awarded (see Table 5.2). As a result, in every year since fiscal year 1998, the EMSP program has chosen to focus all of its new grants in two (or fewer) technical areas (which change from year to year), rather than offering new grants annually in all technical areas.[1] Such discontinuities in funding in specific technical areas from year to year, along with decreases in funding, is not an effective strategy for "expand[ing] the core of 'committed cadre' of investigators who are knowledgeable about EM's problems" (NRC, 1997, p. 4), one of the stated goals of the EMSP program.

In addition, the funding levels of some EQ R&D programs have been

TABLE 5.2 Environmental Management Science Program Funding History (Fiscal Year 1996 to 2000)

|  | FY 1996 | FY 1997 | FY 1998 | FY 1999 | FY 2000 |
|---|---|---|---|---|---|
| Total Budget (in $ millions) | 50.0 | 48.0 | 48.0 | 47.0 | 32.0 |
| Funds available for new starts (in $ millions) | 45.9 | 21.0 | 10.0 | 10.3 | 5.4 |
| Current year funds committed by previous awards ("mortgage", in $ millions) | 0.0 | 23.3 | 34.4 | 31.8 | 22.6 |
| Number of new awards | 136 | 66 | 33 | 39 | |

Source: DOE, 2000m.

---

[1]In fiscal year 1998, EMSP issued two solicitations in the areas of decontamination and decommissioning and high-level waste; in fiscal year 1999, EMSP issued a single solicitation for subsurface contamination and vadose zone issues; in fiscal year 2000, EMSP issued no solicitations for new grants and awarded 31 renewals; and in fiscal year 2001, EMSP issued two solicitations in the areas of deactivation and decommissioning and high-level waste.

developed largely from the "bottom up" and have focused on short-term needs identification (NRC, 1999a), and some have been characterized by significant changes from year to year. For example, a recent NRC committee found that the success of OST's Subsurface Contaminants Focus Area (SCFA)

> has been limited in part by large budget swings. In fiscal year 1998, SCFA's budget was reduced to a level that was insufficient to support significant progress on the development of innovative remediation technologies. The budget level was cut from a 1994 level of $82 million to a 1998 level of $15 million, which included a $5 million congressional earmark, leaving an effective budget of $10 million. This budget was inadequate to fund the types of large-scale demonstrations needed to transition innovative re-mediation technologies from the research and development phase to full-scale application. It also was too small to allow open bidding for project funding. The fiscal year 1999 budget of $25 million, while representing a significant increase, will allow for funding of only a limited number of projects. (NRC, 1999b, p. 247)

The purpose of citing these cases is not to criticize DOE leadership, as a decision to concentrate limited funding in a few high-priority areas to establish critical research foci is a logical alternative to funding only a few projects in all technical areas. Rather, these cases are provided as ex-amples of some of the difficult tradeoffs that must be made when only limited resources are available.

Given the reasonableness of priority on near-term performance, it is likely that the lack of a well-documented, accepted approach for deter-mining long-term R&D funding levels will lead to strong pressure on pro-gram managers to defend and possibly reduce EQ R&D investments.

## DETERMINING AN APPROPRIATE LEVEL OF R&D INVESTMENT

The committee was asked to provide guidance on how to determine the level of future investments in EQ R&D. It has not been possible to identify an analytic or quantitative approach to establish an appropriate level of R&D funding for the EQ business line, because funding levels are in the end a policy decision that involves multiple tradeoffs. However, there are two general techniques that, together, could be used for this purpose: (1) benchmarking against other mission-driven R&D efforts, both nationally and internationally; and (2) applying a set of investment indicators based closely on the adequacy criteria developed in Chapter 3. The committee provides an overview of these techniques in the section that follows. The committee also illustrates by example how each of these techniques could be applied to the EQ R&D portfolio. The committee was not asked to recommend an appropriate level of R&D investment or to recommend that the current level of investment be increased or decreased; however, the committee strongly encourages DOE to conduct its own analyses of EQ R&D funding using these techniques.

## Benchmarking Against Other Mission-Driven R&D Efforts

Benchmarking R&D investment levels with competitors and other similar R&D programs is a standard method used in industry (NRC, 1999a; DOE, 2000j). Benchmarking also can be used to compare the quality and impact of research (as well as the level of R&D investments) in one country with research in other countries, as discussed in reports from the National Academies' Committee on Science, Engineering, and Public Policy (COSEPUP, 1993, 2000). Its recent report, *Experiments in International Benchmarking of U.S. Research Fields* (COSEPUP, 2000), provides a detailed description of the methodology to be used in such benchmarking exercises. There is a large volume of information describing and analyzing R&D funding in the federal government and in the private sector. Although there are marked and understandable differences between DOE's EQ programs and other government and industry R&D programs, benchmarking could provide one meaningful measure for discerning a range of reasonable R&D investment levels for the EQ business line.

The following sections discuss two types of benchmarking and applies them to select EQ R&D funding data: (1) benchmarking total R&D funding and (2) benchmarking the balance of R&D funding by stage of R&D maturity.

## Benchmarking Total R&D Funding

An informative exercise is to compare total EQ R&D funding (both the level of investment and recent funding trends) with that in the three other DOE business lines. As mentioned earlier in this chapter, the National Nuclear Security and Energy Resources business lines both devote approximately 50 percent of their funds to R&D (see Table 5.1). The Science business line, not surprisingly, devotes almost all of its funds to research. Although the current EQ business line has a particular programmatic focus, the 4 percent dedicated to EQ R&D has been called into question by many, including DOE in its department-wide summary of the R&D portfolio effort, *R&D Portfolio Overview,* which states that "current [EQ R&D] funding may not adequately support a long-term integrated research program" (DOE, 2000i, p. 25). The Strategic Laboratory Council's adequacy analysis (DOE, 2000g) and a letter report from the Environmental Management Advisory Board (DOE, 2000h) also have concluded that the level of DOE's EQ R&D funding is inadequate (see Appendix C). Two other groups, the Washington Advisory Group and a National Research Council committee also came to a similar conclusion about the level of funding for subsurface science research in the EQ R&D portfolio (NRC, 2000c; WAG, 1999).

Benchmarking of recent EQ R&D funding trends against funding trends for DOE's other R&D portfolios also is informative because it can help distinguish trends that are DOE-wide from those that are unique to the EQ R&D portfolio. Figure 5.1 shows that EQ R&D funding has declined significantly at the same time that R&D funding for DOE's other business lines has increased. Figure 5.2 illustrates that this reduction is not limited to EM (which dominates total EQ R&D funding data); EQ R&D funding in both RW and NE also has declined significantly in recent years. These data are a strong indication that EQ R&D funding decreases do not simply reflect department-wide (or national) budgetary constraints. DOE, in its *R&D Portfolio Overview*, noted the decline in EQ R&D funding as follows: "The downward funding trend is incongruous

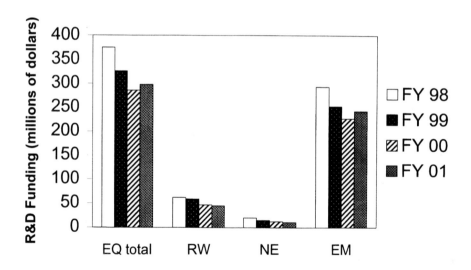

FIGURE 5.2. EQ R&D Spending by Year (in millions of dollars). Data for fiscal year (FY) 1998, 1999, 2000 are from DOE's Environmental Quality Research and Development Portfolio (DOE, 2000b). Data for FY 2001 are from K. Chang, DOE (personal communication).

with upward trends in life-cycle costs and programmatic risk levels associated with current cleanup projects. Further advancements in science and the use of new technologies will be required to meet current cost projections, much less reduce life-cycle costs" (DOE, 2000i, p. 25).

The R&D budgets from other mission agencies also provide a useful comparison. Table 5.1 includes fiscal year 2001 R&D funding data for the U.S. Environmental Protection Agency and the Department of Defense (DOD) in comparison with DOE as a whole, and for DOE's four programmatic R&D business lines. Again, the percentage of the DOE EQ budget spent on R&D is significantly lower than that for other U.S. mission agencies.

The National Science Board's report, *Science and Engineering Indicators 2000* (NSF, 2000) contains extensive data on allocations of R&D funds by federal agencies and the private sector. These data also can provide a context for EQ R&D funding decisions. For example, data on R&D as a percentage of federal budget authority by function is summarized in Table 5.3. These data show that the percentage of DOE's EQ business line budget spent on R&D (4 percent) is significantly lower than the average for the federal government as a whole in the area of natural resources and environment (8.1 percent).

The Industrial Research Institute tracks data on R&D The Industrial Research Institute tracks data on R&D intensity (defined as the ratio of R&D funding to net sales) for different industrial sectors (Table 5.4). Although research intensity is not directly comparable with the percentage of federal program budgets allocated to R&D, the Industrial Research Institute data show that research intensity is highest in knowledge-intensive industries such as software and pharmaceuticals, whereas research intensity is lowest for such mature industries as petroleum and construction. Given the unique and enduring nature of many of DOE's

TABLE 5.3 R&D as a Percentage of Federal Budget Authority by Function

|  | R&D Spending as Percentage of Budget Authority |
| --- | --- |
| General science | 73.1 |
| Space research and technology | 67.3 |
| National defense | 13.4 |
| Agriculture | 10.8 |
| Health | 10.2 |
| Natural resources and environment | 8.1 |
| Transportation | 3.4 |

Source: NSF, 2000; Table 2-2, p. 2-12.

TABLE 5.4 Research and Development Intensity[a] Global Firms (1997)

| Sector | R&D Intensity |
|---|---|
| Software | 13.67 |
| Pharmaceuticals | 12.04 |
| Medical Instruments | 9.67 |
| Scientific Equipment | 6.40 |
| Electronics | 6.30 |
| Computers | 5.96 |
| Chemicals | 4.76 |
| Aerospace | 4.55 |
| Automobile | 4.19 |
| Telecommunications | 3.62 |
| Soaps | 3.55 |
| Heavy Industries | 2.48 |
| Building Materials | 2.04 |
| Food | 1.34 |
| Metal and Metal Products | 1.16 |
| Gas & Electricity | 1.00 |
| Tobacco | 0.95 |
| Forest and Paper Products | 0.90 |
| Engineering and Construction | 0.73 |
| Petroleum | 0.66 |

[a]The ratio of R&D funding to net sales.
Source: Bowonder and Yadav, 1999.

EQ problems (as discussed in Chapter 3), the committee believes that the EQ business line has a fairly knowledge-intensive long-term mission. This is another indication that the EQ R&D budget may be anomalously low with respect to other federal R&D efforts.

**Benchmarking the Balance of R&D Funding**

In theory, benchmarking also can be used to compare the balance of R&D investments in the EQ R&D portfolio with that of other agencies and the private sector. As an example, the committee discusses one element of R&D balance, the percentage of total R&D spending that supports basic research. The National Science Board's report, *Science and Engineering Indicators 2000*, summarizes fiscal year 2000 funding levels for basic research and applied R&D for different types of federal R&D (see Table 5.5). For natural resources and environment, for example, approximately 9 percent of R&D funding supported basic research. For en-

TABLE 5.5 Budget Authority for R&D by Function and Character of Work: Anticipated Levels for Fiscal Year 2000 (millions of dollars)

| Budget Function | Basic Research | Applied Research and Development | R&D Total | Basic Research as a Percentage of R&D Total |
|---|---|---|---|---|
| Total | 18,101 | 57,314 | 75,415 | 24.0 |
| National Defense | 1,152 | 36,559 | 37,710 | 1.7 |
| Nondefense (total) | 16,949 | 20,755 | 37,704 | 45.0 |
| Health | 8,590 | 7,234 | 15,824 | 54.3 |
| Space Research and Technology | 1,841 | 6,581 | 8,422 | 21.9 |
| Energy | 46 | 1,302 | 1,348 | 3.4 |
| General Science | 4,710 | 241 | 4,951 | 95.1 |
| Natural Resources and Environment | 175 | 1,769 | 1,944 | 9.0 |
| Transportation | 634 | 1,206 | 1,840 | 34.5 |
| Agriculture | 736 | 786 | 1,522 | 48.4 |
| All other | 218 | 1,636 | 1,853 | 11.8 |

Source: NSF, 2000, Table 2-3.

ergy, approximately 3 percent of R&D funding supported basic research. The Industrial Research Institute tracks similar data for industry, and its data show that in 2000, approximately 7 percent of industrial R&D funding supported basic research.

Data on the distribution of EQ R&D funding for EM in fiscal year 2000 by focus area, including EMSP funding directed at each focus area (the only significant part of the EQ R&D portfolio with funding data broken down by stage of R&D) are summarized in Table 5.6. The committee has used these data to calculate basic research spending as a percentage of total R&D investment for each of OST's five focus areas, with the following results:

Transuranic and Mixed Waste          4%
Subsurface Contaminants              18%
Tanks                                22%
Deactivation and Decommissioning     28%
Nuclear Materials                    44%

These data show that, with the exception of the Transuranic and Mixed Waste Focus Area, most of OST's focus areas invest a significant

TABLE 5.6 DOE's Office of Science and Technology Fiscal Year 2000 EQ R&D Funding by Stage of Maturity (in $ thousands)

| Focus Area | Basic Research | Applied Research | Exploratory Development | Advanced Development | Engineering Development | Demonstration | Deployment | Total |
|---|---|---|---|---|---|---|---|---|
| Tanks | 13,110 | 4,800 | 3,500 | 4,500 | 12,900 | 10,800 | 8,000 | 57,610 |
| Transuranic and Mixed Waste | 1,228 | 858 | 5,180 | 6,183 | 5,443 | 6,334 | 3,914 | 29,140 |
| Nuclear Materials | 3,189 | 0 | 750 | 950 | 1,295 | 0 | 1,053 | 7,237 |
| Subsurface Contamination | 8,892 | 0 | 0 | 4,500 | 7,300 | 12,200 | 17,300 | 50,192 |
| Deactivation and Decommissioning | 3,598 | 1,646 | 1,646 | 2,171 | 3,264 | 4,608 | 12,671 | 12,672 |

Source: L. Nichols, DOE (personal communication). Basic research funding includes Environmental Management Science Program funding applicable to each focus area.

fraction of their R&D resources in basic research. This probably reflects the significant decrease in total EQ R&D spending over the past few years (see Figure 5.1), coupled with recent congressional pressures to fund basic research within the EMSP program. It should be noted that these data are based on how OST program managers chose to categorize their R&D funding into the 7 stages of R&D tracked by OST, and that this categorization was done relatively quickly at the request of this committee. The committee suspects that some R&D classified as basic research could have been classified as applied research, which would tend to lower the percentages given above. It also is important to recognize that the data do not include relevant basic research funded by DOE's Office of Science, which if incorporated would tend to increase the percentages given above. Due to the significant uncertainties discussed above, the committee cautions the reader not to draw significant conclusions from this comparison. However, the calculation illustrates how DOE could assess the balance of its R&D investments (or at least specific elements of balance) through benchmarking.

**Conclusion**: Benchmarking with other mission-driven federal R&D efforts could provide perspective on whether the EQ R&D budget is too high or too low. It could also help to explain and justify the level of future budget requests to decision makers within DOE, the Office of Management and Budget, and Congress and to other interested parties.

**Recommendation**: DOE should benchmark the EQ R&D budget against other mission-driven federal R&D programs. Such benchmarking exercises should have participation or review by outside experts. Proposed budgets should be presented in the context of benchmarking, and significant deviations from the information gained through benchmarking should be explained.

Such benchmarking should take into account that DOE has a separate basic research program that includes some research activities that are related to (though not directed to) DOE's EQ mission. It also is very important that this analysis to be transparent and credible (COSEPUP, 1999a). Whereas no correct level of investment exists, having a review by internal and external experts can help provide independent advice and enhance credibility in justifying an R&D investment level.

### Indicators of an Adequately Funded R&D Portfolio

In Chapter 3, the committee developed criteria to evaluate the adequacy of the EQ R&D portfolio. These criteria were based on what the committee considered essential elements of a successful, long-term EQ R&D portfolio. And, although they were not framed in terms of R&D investment, they can be re-packaged slightly as investment indicators. The level of EQ R&D investment should be sufficient for the EQ R&D portfolio to:

- address all critical areas of science and technology that are required to address EQ goals and objectives;
- support the accomplishment of closely related DOE and national missions;
- include R&D to develop technical alternatives in cases where (1) existing techniques are expensive, inefficient, or pose high risks to health or the environment; or (2) techniques under development have high technical risk;
- produce results that could transform the understanding, need, and ability to address currently intractable problems and lead to breakthrough technologies;
- leverage R&D conducted by other DOE business lines, the private sector, state and federal agencies, and other nations to address EQ goals and objectives;
- help narrow and bridge the gap between R&D and application in the field;
- improve performance, reduce risks to human health and the environment, decrease cost, and advance schedules.
- achieve an appropriate balance between addressing long-term and short-term issues;
- involve a diversity of participants from academia, national laboratories, other federal agencies, and the private sector, including students, postdoctoral associates, and other early-career researchers;
- include annual new starts, extensions of promising R&D, and periodic new initiatives.

Meeting such criteria is an important indication of an appropriately formulated R&D portfolio. Although the level of R&D investment alone cannot guarantee the achievement of these indicators, the level of funding should not preclude their achievement.

**Finding: It has not been possible to identify an analytic or quantitative approach that is suitable for establishing an appropriate level of R&D funding for the EQ R&D portfolio.**

**Conclusion: Investment indicators based on the functions of a successful EQ R&D portfolio can provide useful guides for the appropriate funding level.**

**Recommendation: DOE should use investment indicators, together with benchmarking techniques, to help determine the appropriate level of EQ R&D investments.**

These investment indicators should provide a useful guide to the appropriate range of EQ R&D funding levels. As discussed in Chapter 3, it is also particularly important that DOE's process for arriving at appropriate level of R&D funding consider the contributions of EQ R&D to meeting DOE's other missions (particularly, its National Nuclear Security and Energy Resources missions), and should be based on a retrospective examination of the results of past EQ R&D.

## CONCLUSION

DOE's EQ R&D portfolio must be recognized as centrally important to DOE's EQ and other missions and as an enduring responsibility of the department. R&D success requires an adequate, stable, and predictable level of funding. A well-designed, sufficiently funded, and well-implemented EQ R&D portfolio is necessary, but not sufficient, to assure that the potential value of R&D in addressing DOE's EQ problems is achieved. Many other features also must be present, including technically competent and trusted R&D program managers; effective relationships among problem holders, R&D managers and researchers; good communication of R&D results; and incentives for R&D results to be used in solving problems.

An effective portfolio also requires close and trusting relationships among the responsible DOE headquarters and local officials, contractors at the sites, state regulatory officials, and stakeholders such as the affected community. The nature of successful EQ R&D is to present opportunities to reduce risks to workers and the public, improve schedules, decrease costs, and solve problems (see discussion in Chapter 3). But it also can require re-addressing existing agreements, changing schedules, dealing with periods of uncertainty, and revisiting expectations. All of these factors must be resolved for DOE's EQ R&D to achieve its goals. An EQ R&D portfolio that is well conceived, effectively managed, adequately and consistently funded, and championed by DOE leadership is essential to success in achieving the DOE EQ mission.

# REFERENCES

AAAS (American Association for the Advancement of Science). 2001. Congressional Action on Research and Development in the FY 2001 Budget. Washington, D.C.: AAAS.

Bjornstadt, D. J., D. W. Jones, M. Russel, K. S. Redus, and C. L. Dummer. 1998. Outcome-Oriented Risk Planning for DOE's Clean Up. JIEE 98-01. Knoxville, Tenn.: Joint Institute for Energy and the Environment.

Bjornstadt, D.J, D. W. Jones, C. L. Dummer, and K. S. Redus. 2000. Investment-Oriented R&D Planning and Evaluation for DOE's Cleanup. JIEE 2000-02. Knoxville, Tenn.: Joint Institute for Energy and Environment.

Bowonder, B. and S. Yadav. 1999. R&D Spending Patterns of Global Firms. Research-Technology Management. November-December: 44-55. Washington, D.C.: Industrial Research Institute.

COSEPUP (Committee on Science, Engineering, and Public Policy). 1993. Science, Technology, and the Government: National Goals for a New Era. Washington, D.C.: National Academy Press.

COSEPUP. 1999a. Capitalizing on Investments in Science and Technology. Washington, D.C.: National Academy Press.

COSEPUP. 1999b. Evaluating Federal Research Programs: Research and the Government Performance and Results Act. Washington, D.C.: National Academy Press.

COSEPUP. 2000. Experiments in International Benchmarking of U.S. Research Fields. Washington, D.C.: National Academy Press.

CRESP (Consortium for Risk Evaluation with Stakeholder Participation). 1999. Peer Review of the U.S. Department of Energy's Use of Risk in its Prioritization Process. December 15. Washington, D.C.: U.S. Department of Energy.

DOE (U.S. Department of Energy). 1995. Closing the Circle on the Splitting of the Atom. DOE/EM-0266 (reprinted 1996). Washington, D.C.: Government Printing Office.

DOE. 1996. Taking Stock: A Look at the Opportunities and Challenges Posed by Inventories From the Cold War Era. Washington, D.C.: U.S. Department of Energy. 19.

DOE. 1997a. Linking Legacies: Connecting the Cold War Nuclear Weapons Production Processes to Their Environmental Consequences. DOE/EM-0319. Washington, D.C.: U.S. Department of Energy.

DOE. 1997b. U.S. Department of Energy Strategic Plan: Proving America with Energy Security, National Security, Environmental Quality, Science Leadership. DOE/PO-0053. September Washington, D.C.: U.S. Department of Energy.

DOE. 1998. Accelerating Cleanup: Paths to Closure. DOE/EM-0362. June. Washington, D.C.: U.S. Department of Energy/Office of Environmental Management.

DOE. 1999a. From Cleanup to Stewardship: A Companion report to *Accelerating Cleanup: Paths to Closure.* DOE/EM-0466. Washington, D.C.: U.S. Department of Energy.

DOE. 1999b. Resolution on the Work Package Ranking System. Technology Development and Transfer Committee of Department of Energy Environmental Management Advisory Board. April 22. Washington, D.C.: U.S. Department of Energy.

DOE. 2000a. Energy Resources Research and Development Portfolio. Volume 1 of 4, February. Washington, D.C.: U.S. Department of Energy.

DOE. 2000b. Environmental Quality Research and Development Portfolio. Volume 2 of 4, February. Washington, D.C.: U.S. Department of Energy.

DOE. 2000c. National Security Research and Development Portfolio. Volume 3 of 4, February. Washington, D.C.: U.S. Department of Energy.

DOE. 2000d. Science Research and Development Portfolio. Volume 4 of 4, February. Washington, D.C.: U.S. Department of Energy.

DOE. 2000e. Status Report on Paths to Closure. DOE/EM-0526. March. Washington, D.C.: U.S. Department of Energy.

DOE. 2000f. Strategic Plan: Strength Through Science, Powering the 21st Century. DOE/CR-0070. September. Washington, D.C.: U.S. Department of Energy.

DOE. 2000g. Adequacy Analysis of the Environmental Quality Research and Development Portfolio. September. Washington, D.C.: U.S. Department of Energy.

DOE. 2000h. Environmental Management Advisory Board Letter Report on Adequacy Analysis. October. Washington, D.C.: U.S. Department of Energy.

DOE. 2000i. R&D Portfolio Analysis, Overview. Online. Available at: http://www.osti.gov/portfolio.

DOE. 2000j. Review of the Department of Energy's Laboratory-Directed Research and Development Program. External Members of the DOE Laboratory Operations Board. Washington, D.C.: U.S. Department of Energy.

DOE. 2000k. Long-Term Nuclear Technology Research and Development Plan. Nuclear Energy Research Advisory Committee, Subcommittee on Long-Term Planning for Nuclear Energy Research. Washington, D.C.: U.S. Department of Energy.

DOE. 2000l. Report of the Secretary of Energy Advisory Board's Panel on Emerging Technological Alternatives to Incineration. Washington, D.C.: U.S. Department of Energy.

DOE. 2000m. Environmental Management Science Program Fiscal Year 2000-2004 Multi-Year Program Plan. DOE/ID-10730. Washington, D.C.: U.S. Department of Energy.

DOE. 2000n. The Office of Science and Technology Work Package Ranking System. March. Washington, D.C.: U.S. Department of Energy.

DOE. 2001a. The Role and Status of Basic Science in Accomplishing the Department of Energy's Environmental Management Mission. Environmental Management Advisory Board. Washington, D.C.: U.S. Department of Energy.

DOE. 2001b. A Report to Congress on Long-Term Stewardship. DOE/EM-0563. Washington, D.C.: U.S. Department of Energy.

Fischoff, B. 2000. Scientific management of science? Policy Sciences 33(1):73-87.

GAO (U.S. General Accounting Office). 1996. Energy Management: Technology Development Program Taking Action to Address Problems. GAO/RCED-96-184. July. Washington, D.C.: U.S. General Accounting Office.

GAO. 1998. Further Actions Needed to Increase the Use of Innovative Cleanup Technologies. GAO/RCED-98-249. September. Washington, D.C.: U.S. General Accounting Office.

IIASA (International Institute for Applied Systems Analysis/World Energy Congress). 1995. Global Energy Perspectives to 2050 and Beyond. Laxenburg, Austria, 106 pp.

Jenni, K. E., M. W. Merkofer, and C. Williams. 1995. The rise and fall of a risk-based priority system: Lessons Learned from DOE's Environmental Restoration Priority System. Risk Analysis 15(3):397-410.

Kaiser-Hill. 2000. Rocky Flats Environmental Technology Site Programmatic Risk Management Plan: A Plan to Identify, Prioritize, and Mitigate Risks to Project Schedule and Cost. Denver, Colo.: Kaiser-Hill Company, L.L.C.

NRC (National Research Council). 1990. Rethinking High-Level Radioactive Waste Disposal. Washington, D.C.: National Academy Press.

NRC. 1995. Nuclear Wastes: Technologies for Separations and Transmutation. Washington, D.C.: National Academy Press.

NRC. 1996a. The Waste Isolation Pilot Plant: A Potential Solution for the Disposal of Transuranic Waste. Washington, D.C.: National Academy Press.

NRC. 1996b. The Hanford Tanks: Environmental Impacts and Policy Choices. Washington, D.C.: National Academy Press.

NRC. 1997. Building an Effective Environmental Management Science Program: Final Assessment. Washington, D.C.: National Academy Press.

NRC. 1998. Peer Review in Environmental Technology Development Programs: The Department of Energy's Office of Science and Technology. Washington, D.C.: National Academy Press.

NRC. 1999a. Decision Making in the U.S. Department of Energy's Environmental Management Office of Science and Technology. Washington, D.C.: National Academy Press.

NRC. 1999b. Groundwater and Soil Cleanup: Improving Management of Persistent Contaminants. Washington, D.C.: National Academy Press.

NRC. 1999c. An End State Methodology for Identifying Technology Needs for Environmental Management, with an Example from the Hanford Site Tanks. Washington, D.C.: National Academy Press.

NRC. 1999d. Alternative High-Level Waste Treatments at the Idaho National Engineering and Environmental Laboratory. Washington, D.C.: National Academy Press.

NRC. 2000a. Long-Term Institutional Management of U.S. Department of Energy Legacy Waste Sites. Washington, D.C.: National Academy Press.

NRC. 2000b. From Research to Operations in Weather Satellites and Numerical Weather Prediction. Washington, D.C.: National Academy Press.

NRC. 2000c. Research Needs in Subsurface Science: U.S. Department of Energy's Environmental Management Science Program. Washington, D.C.: National Academy Press.

NRC. 2000d. Long-Term Research Needs on Radioactive High-Level Waste at Department of Energy Sites: Interim Report. Washington, D.C.: National Academy Press.

NRC. 2000e. Long-Term Research Needs in Deactivation and Decommissioning at Department of Energy Sites: Interim Report. Washington, D.C.: National Academy Press.

NRC. 2000f. Alternatives for High-Level Waste Salt Processing at the Savannah River Site. Washington, D.C.: National Academy Press.

NRC. 2000g. Alternative High-Level Waste Treatments at the Idaho National Engineering and Environmental Laboratory. Washington, D.C.: National Academy Press.

NRC. 2000h. Natural Attenuation for Groundwater Remediation. Washington, D.C.: National Academy Press.

NRC. 2000i. Improving Operations and Long-Term Safety of the Waste Isolation Pilot Plant: Interim Report. Washington, D.C.: National Academy Press.

NRC. 2000j. Grand Challenges in Environmental Sciences. Washington, D.C.: National Academy Press.

NRC. 2001. Disposition of High-Level Waste and Spent Nuclear Fuel: The Continuing Societal and Technical Challenges. Washington, D.C.: National Academy Press.

NSF (National Science Foundation). 2000. Science and Engineering Indicators 2000. Washington, D.C.: National Science Foundation.

NWTRB (Nuclear Waste Technical Review Board). 1999. NWTRB Viewpoint Fact Sheet: *What is the Nuclear Waste Technical Review Board?* January.

NWTRB. 2000. Letter Report to the U.S. Congress and the Secretary of Energy. December. Washington, D.C.

OTA (Office of Technology Assessment). 1991. Complex Cleanup: The Environmental Legacy of Nuclear Weapons Production, OTA-O-484. Washington, D.C.: U.S. Government Printing Office.

OTA. 1993a. Dismantling the Bomb and Managing the Nuclear Materials, 1-16. Washington, D.C.: U.S. Government Printing Office.

OTA. 1993b. Hazards Ahead: Managing Cleanup Worker Health and Safety at the Nuclear Weapons Complex, 1-3. Washington, D.C.: U.S. Government Printing Office.

Probst, K. N. and McGovern, M.H. 1998. Long-Term Stewardship and the Nuclear Weapons Complex: The Challenge Ahead. Center for Risk Assessment, Resources for the Future, Washington, D.C.

U.S. House of Representatives. 2000. Incinerating Cash: The Department of Energy's Failure to Develop and Use Innovative Technologies to Clean Up the Nuclear Waste Legacy. October. Washington, D.C.: U.S. House of Representatives.

WAG (Washington Advisory Group). 1999. Management Subsurface Contamination: Improving Management of the Department of Energy's Science and Engineering Research on Subsurface Contamination. December 17. Washington D.C.

White House. 1999. Strategic Plan to Implement Executive Order 13101: Greening the Government Through Waste Prevention, Recycling, and Federal Acquisition. March 12. Washington, D.C.

# APPENDIX A

# BIOGRAPHICAL SKETCHES OF COMMITTEE MEMBERS

**Gregory R. Choppin**, *Chair*, is currently the R.O. Lawton Distinguished Professor of Chemistry at Florida State University. His research interests involve the chemistry of the f-elements, the separation science of the f-elements, and the physical chemistry of concentrated electrolyte solutions. During a postdoctoral period at the Lawrence Radiation Laboratory, University of California, Berkeley, he participated in the discovery of mendelevium, element 101. His research and educational activities have been recognized by the American Chemical Society's Award in Nuclear Chemistry, the Southern Chemist Award of the American Chemical Society, the Manufacturing Chemist Award in Chemical Education, the Chemical Pioneer Award of the American Institute of Chemistry, a Presidential Citation Award of the American Nuclear Society, the Bequerel Award in Nuclear Chemistry of the British Royal Society of Chemistry, and honorary D.Sc. degrees from Loyola University and the Chalmers University of Technology (Sweden). Dr. Choppin has served on the National Research Council's (NRC's) Board on Chemical Sciences and Technology. He received a B.S. degree in chemistry from Loyola University, New Orleans, and a Ph.D. degree from the University of Texas, Austin.

**David E. Adelman** serves as staff attorney for the international and nuclear programs at the Natural Resources Defense Council. Prior to joining the Council in 1998, Dr. Adelman served as an associate with the law firm of Covington & Burling, focusing on environmental and intellectual property litigation and regulatory matters. He is a member of the Department of Energy's (DOE's) Environmental Management Advisory Board. Dr. Adelman received his B.A. in chemistry and physics from Reed College, his Ph.D. in chemical physics from Stanford University, and his J.D. from Stanford Law School.

**Radford Byerly, Jr.** retired as vice-president for public policy of the University Corporation for Atmospheric Research after a distinguished career in academia and government, specializing in science management and policy. Dr. Byerly is the co-author of several recent papers on federal research and development (R&D) policy, including "Beyond Basic and Applied" (Physics Today, 1998) and "The Changing Ecology of United States Science" (Science, 1995). Among his many positions, Dr. Byerly has worked at the National Institute of Standards and Technology (then the National Bureau of Standards) in the environmental measurement and fire research programs; has served as chief of staff of the U.S. House of Representatives Committee on Science and Technology; and was director of the University of Colorado's Center for Space and Geosciences Policy. He currently serves on the American Association for the Advancement of Science's Committee on Science, Engineering, and Public Policy and serves on National Science Foundation site visit committees and review panels. He is a member of the Board of Associated Universities for Research in Astronomy, and has served on the Committee on the Department of Energy—Office of Science and Technology's Peer Review Program. He received his Ph.D. in physics from Rice University.

**William L. Friend** is a corporate director, consultant, and educator drawing on his background of over 40 years in chemical engineering and executive management in the international engineering-construction industry. He recently retired as executive vice president of the Bechtel Group. During his tenure there, he was responsible for Bechtel's DOE/Department of Defense/National Aeronautics and Space Administration activity, including environmental remediation work at the Hanford reservation and other DOE sites. He currently chairs the University of California President's Council for the National Labs. Mr. Friend was elected to the National Academy of Engineering for leadership in the development of new technologies and their application in commercial facilities. He received his Bachelors Degree from Polytechnic University and holds a Masters in Chemical Engineering from the University of Delaware.

**Thomas H. Isaacs** is director of Lawrence Livermore National Laboratory's Office of Policy, Planning, and Special Studies and chair of its Council on Energy and Environmental Systems. Mr. Isaacs is responsible for long-range strategic and institutional planning and conducts policy and technology studies for the laboratory. Prior to joining the laboratory in 1996, he held various positions within DOE, including executive director of DOE's Advisory Committee on External Regulation of DOE Nuclear Safety and Director of Strategic Planning and International Programs for the Department's Radioactive Waste

Program. Mr. Isaacs received a B.S. in chemical engineering from the University of Pennsylvania and a M.S. in engineering and applied physics from Harvard University.

**James H. Johnson, Jr.** is professor and dean of the College of Engineering, Architecture, and Computer Sciences at Howard University. Dr. Johnson's research interests have focused mainly on the reuse of wastewater treatment sludges and the treatment of hazardous substances. His research has included the refinement of composting technology for the treatment of contaminated soils, chemical oxidation and cometabolic transformation of explosive-contaminated wastes, biodegradation of fuel-contaminated groundwater, the evaluation of environmental policy issues in relation to minorities and development of environmental curricula. Currently, he serves as associate director of the Great Lakes and Mid-Atlantic Center for Hazardous Substance Research and as a member of the Environmental Protection Agency's (EPA's) Office of Research and Development, Board of Science Counselors and the NRC's Board on Radioactive Waste Management. He has served on the Environmental Engineering Committee of EPA's Science Advisory Board and NRC's Committee on the Remediation of Buried and Tank Wastes. Dr. Johnson is a registered professional engineer in the District of Columbia, a Diplomate in the American Academy of Environmental Engineers and a fellow of the American Society of Civil Engineers. He received a B.S. from Howard University, a M.S. from University of Illinois, and a Ph.D. from the University of Delaware.

**Charles Kolstad** is a professor of environmental economics at the University of California, Santa Barbara, where he is jointly appointed in the Department of Economics, the Ben School of Environmental Science and Management, and the Environmental Studies Program. Dr. Kolstad's current research focuses on the role of information in environmental decision making and regulation, and environmental valuation theory. He also has a major research project on the role of uncertainty and learning in controlling the precursors of climate change. His past work on energy markets has focused on coal and electricity markets, including the effect of air pollution regulation on these markets. Dr. Kolstad is president-elect of the Association of Environmental and Resource Economics and editor of the journal *Resource and Energy Economics*. He is also a member of EPA's Clean Air Act Compliance Analysis Committee, and has served on numerous other advisory boards, including the Environmental Economics Advisory Committee of EPA's Science Advisory Board. Dr. Kolstad has served on the NRC's Board on Energy and Environmental Systems, the Energy Engineering Board, and the Committee on Fuel Economy of Automobiles and Light Trucks. He received his B.S. from Bates College, his M.A. from University of Rochester, and his Ph.D. from Stanford University.

**C. Edward Lorenz** recently retired as vice president of research and development for DuPont Chemicals, of E.I. DuPont de Nemours & Co. Dr. Lorenz began his career at DuPont as a research chemist, and held a variety of research and management positions throughout his four-decade career with the firm. He holds several patents in the fields of catalysis, monomer, and polymer synthesis. Dr. Lorenz is a member of the American Chemical Society, the Society of the Chemical Industry, the New York Academy of Sciences, and the American Association for the Advancement of Science. He has served on Industry Advisory Committees for New York University, the University of Georgia, and the University of Tennessee. Dr. Lorenz received his B.S. and Ph.D. degrees in organic chemistry from New York University.

**Michael Menke** is a consultant at Hewlett-Packard. Prior to joining Hewlett-Packard, Dr. Menke was president of Value Creation Associates where he worked with research-driven companies in developing successful business and technology strategies, re-engineering their R&D management and new product development processes, and improving R&D productivity. He was a founding partner of Strategic Decisions Group and led its R&D and pharmaceutical industry practices, as well as its groundbreaking benchmark study of the best decision practices of the world's leading companies. Dr. Menke has published extensively and speaks frequently on a wide range of business and innovation management topics. His consulting assignments include new product commercialization strategies, product sales forecasting and capacity planning, R&D portfolio management, and evaluation of new high-technology products and processes in a wide range of industries, including biotechnology, chemicals, medical devices, and pharmaceuticals. Dr. Menke has served on the NRC's Committee on Prioritization and Decision Making in the U.S. Department of Energy Office of Science and Technology. He received a B.A. in physics from Princeton, a M.Sc. in applied math from Cambridge, and a Ph.D. in physics from Stanford University.

**Warren F. Miller** recently retired from his position as a senior advisor to the director of Los Alamos National Laboratory and professor-in-residence in the Department of Nuclear Engineering at the University of California, Berkeley. Dr. Miller has extensive experience in the area of R&D program management. He served in a variety of management positions at Los Alamos National Laboratory, including deputy director for science and technology (1996-1999), director of science and technology base programs (1993-1995), associate laboratory director for research and education (1992-1993), deputy laboratory director (1986-1988), associate laboratory director for energy programs (1981-1982), and as deputy associate laboratory director for nuclear programs (1980-1981). He also served as the E.H. and M.E. Pardee Professor in the Department of Nuclear Engineering at the University of California, Berkeley, from 1988 to 1992. Dr. Miller received a B.S. in engineering science from the U.S. Military Academy and his M.S. and Ph.D. degrees in nuclear engineering from Northwestern University. Dr. Miller was elected to the National Academy of Engineering in 1996.

**Victoria Tschinkel** is senior consultant for environmental issues at the law firm of Landers and Parsons, Tallahassee, Florida. In this position, she specializes in assisting corporate clients on strategic environmental issues and represents clients before agencies and the Legislature. Ms. Tschinkel served as Secretary of the Florida Department of Environmental Regulation (1981-1987) and has held positions on a number of national advisory councils such as the National Environmental Enforcement Council and the Energy Research Advisory Board. She is a member of the National Academy of Public Administration and continues to serve as a member of both state and national advisory councils. She is a director of Phillips Petroleum Company, Resources for the Future, and the Center for Clean Air Quality. She currently serves as a member of the NRC's Board on Radioactive Waste Management, and is a former member of the Commission on Geosciences, Environment, and Resources. Ms. Tschinkel has served on numerous NRC study committees, including the Committee to Evaluate the Science, Engineering, and Health Basis of the Department of Energy's Environmental Management Program, the Committee on Remedial Action Priorities for Hazardous Waste Sites, and the Committee to Provide Interim Oversight of the DOE Nuclear Weapons Complex. Ms. Tschinkel received her B.S. degree in zoology from the University of California, Berkeley.

# APPENDIX B

# PARTICIPANTS LIST AND AGENDA FOR AUGUST WORKSHOP

## COMMITTEE MEMBERS

Gregory R. Choppin, *Chair*, Florida State University, Tallahassee
Teresa Fryberger,[1] *Vice-Chair*, Brookhaven National Laboratory, Upton, New York
David E. Adelman, Natural Resources Defense Council, Washington, D.C.
Radford Byerly, Jr., Independent Consultant, Boulder, Colorado
William L. Friend, Bechtel Group Inc. (retired), Mclean, Virginia
Thomas Isaacs, Lawrence Livermore National Laboratory, Livermore, California
James H. Johnson, Jr., Howard University, Washington, D.C.
Charles Kolstad, University of California, Santa Barbara
C. Edward Lorenz, E.I. Dupont De Nemours & Co. (retired), Wilmington, Delaware
Michael Menke, Hewlett-Packard, Palo Alto, California
Warren Miller, Jr., Los Alamos National Laboratory (retired), New Mexico
Victoria Tschinkel, Landers and Parsons, Tallahassee, Florida

## SPEAKERS

John Gibbons, Office of Science and Technology Policy (retired), The Plains, Virginia
Ivan Itkin, U.S. Department of Energy, Washington, D.C.

---

[1] Recused herself from committee activities in November 2000 and resigned from committee in January 2001 after accepting a management position within the Department of Energy Office of Environmental Management.

David Heyman, U.S. Department of Energy, Washington, D.C.
James Owendoff, U.S. Department of Energy, Washington, D.C.


## OTHER PARTICIPANTS

John Applegate, Indiana University School of Law, Bloomington
Robert Bernero, U.S. Nuclear Regulatory Agency (retired), Gaithersburg, Maryland
Paul Bertsch, University of Georgia Savannah River Ecology Laboratory, Aiken, South Carolina
David Bjornstad, Oak Ridge National Laboratory, Tennessee
Margaret Cavanaugh, National Science Foundation, Arlington, Virginia
Ker-Chi Chang, U.S. Department of Energy, Washington, D.C.
Thomas Cotton, JK Research Associates, Inc., Washington, D.C.
Allen Croff, Oak Ridge National Laboratory, Tennessee
Thomas Essig, Nuclear Regulatory Commission, Rockville, Maryland
Stephen Goldberg, U.S. Department of Energy, Washington, D.C.
Norman Edelstein, U.S. Department of Energy, Washington, D.C.
Greg Frandsen, Bechtel BWXT Idaho, LLC, Idaho Falls, Idaho
Mark Gilbertson, U.S. Department of Energy, Washington, D.C.
Paul Hart, U.S. Department of Energy, Morgantown, West Virginia
James Helt, Argonne National Laboratory, Illinois
Thomas Hirons, Los Alamos National Laboratory, New Mexico
Paul Kearns, Bechtel BWXT Idaho, LLC, Idaho Falls, Idaho
Mack Lankford, U.S. Department of Energy, Germantown, Maryland
Stephen Lingle, U.S. Environmental Protection Agency, Washington, D.C.
David Maloney, Kaiser-Hill Company, Golden, Colorado
Robert Marianelli, Office of Science and Technology Policy, Washington, D.C.
Lana Nichols, U.S. Department of Energy, Germantown, Maryland
Timothy Oppelt, U.S. Environmental Protection Agency, Washington, D.C.
William Ott, U.S. Nuclear Regulatory Commission, Rockville, Maryland
Trueman Parish, Kingsport, Tennessee
John Pendergrass, Environmental Law Institute, Washington, D.C.
Loni Peurrung, Battelle-Albuquerque, New Mexico
Mike Pfister, Coleman Federal, Fairfax, Virginia
Rod Quinn, Battelle-Albuquerque, New Mexico
Paul Smith, U.S. Department of Energy, Germantown, Maryland
Gary Tritle, Lakeway, Texas
John Veldman, Westinghouse Savannah River Company, Aiken, South

Carolina
Jef Walker, U.S. Department of Energy, Germantown, Maryland
C. Herb Ward, Rice University, Houston, Texas
Susan Wiltshire, Hamilton, Massachusetts
Nicholas Woodward, U.S. Department of Energy, Germantown,
    Maryland
Raymond Wymer, Oak Ridge, Tennessee

# Building a Long-Term Environmental Quality Research and Development Program in the U.S. Department of Energy

## Workshop – August 23, 24 & 25, 2000

### *National Academies*
### *National Academy of Sciences Building*
### *2101 Constitution Avenue*
### *Washington, DC 20418*

## PUBLIC AGENDA

### Wednesday, August 23

| OPEN SESSION (Committee, Guests, and NRC Staff) – Lecture Room |
| --- |

8:30 a.m.       Welcome and Introductions
                *Gregory Choppin*, Chair
                *Teresa Fryberger,* Vice-Chair

8:45 a.m.       Overview of DOE's R&D Portfolio Process and Study
                *David Heyman*, Senior Advisor to the Secretary,
                Technology, Policies, and Partnerships.

9:15 a.m.       Discussion

9:30 a.m.       Office of Environmental Management's Long-Term
                Environmental Quality R&D Needs
                *James Owendoff*, Principal Deputy Assistant
                Secretary for Environmental Management

9:50 a.m.       Discussion

10:00 a.m.      Office of Civilian Radioactive Waste Management's Long-Term
                Environmental Quality R&D Needs
                *Ivan Itkin*, Director of Office of Civilian Radioactive
                Waste Management

10:15 a.m.      Discussion

10:30 a.m.      BREAK

| 10:45 a.m. | Evaluating the Benefits of Long-Term Environmental Quality R&D |
| | *Jack Gibbons*, Former Assistant to the President for Science and Technology, and Former Director, OSTP |

| 11:00 a.m. | Discussion |

| 11:15 a.m. Working | Introduction to R&D Portfolio Analysis and Overview of |
| | Group Tasks |
| | *Michael Menke*, Committee Member |

| 11:45 a.m. | Discussion |

| 12:00 noon | LUNCH available in the Refectory. |

---

**NOTE: Working Group Leads, Chair and Vice-Chair Meet for Lunch in CLOSED SESSION (Committee and NRC Staff ONLY) – Room 150**

---

| 1:00 p.m. | Break into Working Groups A, B, and C for 1-hour Discussion of Morning Presentations and Working Group Charges (Lecture Room, Room 150, Room 180) |

| 2:00 p.m. | Brief Plenary Discussion of Working Group Discussions (10 minute reports/10 minute discussion for each working group) |

| 2:30 p.m. | Break into Working Groups for First Working Session (see description of working group tasks) |

| 4:45 p.m. | Working Groups Report Back to Plenary Session |

| 5:45 p.m. | Adjourn |

| 6:00 p.m. | Reception (Great Hall) |

## Thursday, August 24

---

**CLOSED SESSION (Committee and NRC Staff ONLY) – Room 150**

---

| 7:30 a.m. | EQ Committee Meets for Breakfast |

---

**OPEN SESSION (Committee, Guests, and NRC Staff) – Lecture Room**

---

8:30 a.m.          Plenary Session: Overview/instructions
                              *Gregory Choppin, Chair*
                              *Teresa Fryberger, Vice-Chair*

8:45 a.m.          Break Into Working Groups

11:00 a.m.        Working Groups Report Back to Plenary

12:00 noon       LUNCH available in the Refectory

1:00 p.m.          Plenary Session: General Instructions for Final Working Group
                        Sessions
                              *Gregory Choppin, Chair*
                              *Teresa Fryberger, Vice-Chair*

1:30 p.m.          Working Groups Meet for Last Time

3:30 p.m.          Working Groups Report Back to Plenary

4:30 p.m.          Adjourn Open Session

| **CLOSED SESSION (Committee and NRC Staff ONLY)** |
| --- |

6:00 p.m.          **Committee Dinner (Executive Dining Room)**

## Friday, August 25

| **CLOSED SESSION (Committee and NRC Staff ONLY) – NAS Board Room** |
| --- |

8:30 a.m.          **Commence Meeting**

4:00 p.m.          **Adjourn**

# APPENDIX C

# SUMMARY OF PREVIOUS REVIEWS OF DOE'S ENVIRONMENTAL QUALITY RESEARCH AND DEVELOPMENT PORTFOLIO

The Department of Energy's (DOE's) Strategic Laboratory Council (SLC) recently conducted an analysis to determine the adequacy of the current portfolio of DOE research and development (R&D) activities to meet the objectives of the Environmental Quality (EQ) business line (DOE, 2000g). After the SLC's analysis was published, the Technology Development and Transfer Committee of DOE's Environmental Management Advisory Board (EMAB) commented on the results of the analysis, evaluated the process used to develop the analysis, and offered recommendations in a letter report (DOE, 2000h).

The major findings and recommendations from the adequacy analyses are summarized below, followed by a table summarizing the identified major EQ R&D gaps and opportunities (Table C.1). The full text of the EMAB letter report is included at the end of this appendix.

### Adequacy Analysis of the Environmental Quality Research and Development Portfolio (DOE, 2000g)

The SLC panel arrived at the following conclusions:

The EQ R&D Portfolio adequately addressed three of the ten technology categories:

- manage mixed low-level and TRU wastes;
- manage spent nuclear fuel; and
- dispose high-level waste, spent nuclear fuel, and nuclear materials.

Three of the ten technology categories were addressed in a moderately adequate manner:

- manage high-level waste;
- manage nuclear material; and
- dispose TRU, low-level, mixed low-level, and hazardous waste.

Four of the ten technology categories were inadequately addressed:

- environmental remediation;
- deactivate and decommission;
- minimize waste generation; and
- long-term stewardship.

The panel considered the magnitude of the gaps for each technology category and how important filling those gaps is to meeting the EQ strategy and objectives. The panel combined these estimates of the significance of these gaps with the adequacy assessment to identify four priority areas for improving the portfolio:

- environmental restoration;
- manage high-level waste;
- deactivate and decommission; and
- long-term stewardship.

The SLC panel developed a number of findings and recommendations on how DOE might improve its EQ R&D portfolio:

**Finding 1**: The EQ Portfolio has significant gaps and, as a whole, is underinvested.

**Recommendation 1**: Additional R&D funding is warranted for priority investments. The highest priority areas are: environmental restoration; manage high-level waste; deactivation and decommissioning; and long-term stewardship.

**Finding 2**: The R&D portfolio does not include a longer-term vision and "strategic" elements such as alternative technologies and next-generation solutions.

**Recommendation 2**: Part of the R&D portfolio needs to focus on the long-term mission to provide fundamental information that will allow for better understanding and definition of the larger, more difficult problems that will not be solved in the next 5 to 10 years. A portion of the R&D profile should be devoted to strategic R&D, such as "backup" technologies in high risk/high budget areas to reduce the programmatic risk to the department.

**Finding 3**: The funding distribution across the maturity spectrum is unbalanced.

**Recommendation 3**: The portfolio needs to be more balanced across the technical maturity spectrum without sacrificing recent successes in technology deployment. The maximum benefit from R&D will be obtained through a balanced portfolio that will foster the development of next generation solutions from basic R&D through applied research and development and ultimately to deployment. Basic research should continue to be targeted at a broad spectrum of disciplines that are relevant to the issues facing the EQ business line. Important areas of investment in applied research include separations, robotics, characterization and sensors, and institutional controls related to stewardship.

**Finding 4**: Significant life-cycle costs and corresponding R&D hinge on highly uncertain end states.

**Recommendation 4**: DOE must continue to emphasize the development of waste acceptance criteria and definition of end states for both sites and facilities. This includes the need to gather data and develop fundamental knowledge that supports these efforts.

**Finding 5**: Additional effort is required to identify priorities based on risk.

**Recommendation 5a**: DOE must develop a better understanding of the risk associated with hazardous materials and develop tools that credibly represent those risks in an open and transparent manner in order to increase the ability to balance human health and environmental risk with other considerations in DOE decision making.

**Recommendation 5b**: DOE must develop a better understanding of programmatic risks and their potential impact on meeting DOE objectives to improve the long-term management of EQ problems. This supports recommendation 2 on the need for alternative approaches in high risk/high cost areas.

**Finding 6**: Technology Categories are highly interdependent.

**Recommendation 6**: Both "Long-Term Stewardship" and "Minimize Waste Generation" categories require additional emphasis and the associated R&D should be applied across the other EQ objectives.

**Finding 7**: Interfaces among business lines are not adequate to establish fully complementary and synergistic programs.

**Recommendation 7a**: Interfaces with other DOE business lines and their portfolios should continue to be recognized, developed, and fostered. Synergism and exchange of information should be sought out and acknowledged where appropriate.

**Recommendation 7b**: Continue to improve the portfolio process so that it will provide a long-term view of the DOE business lines.

TABLE C.1 R&D Gaps and Opportunities Identified by the Strategic Laboratory Council's Adequacy Analysis

| Technical Category | Gaps | Opportunities |
|---|---|---|
| Environmental Remediation | • Understanding of fate and transport of contaminants in the vadose zone (including improved characterization and modeling of contamination in the vadose zone).<br>• Sensors and characterization technologies particularly for dense non-aqueous liquids (DNAPLs) but also for metal and radioactive contaminants.<br>• Understanding of long-term performance of in-situ containment and stabilization techniques, including caps.<br>• Understanding of natural attenuation and the potential role of monitored natural attenuation as a complement or alternative to active remediation. | • Develop improved understanding and modeling of vadose zone fate and transport processes to support improved risk analysis and remediation and monitoring system designs.<br>• Develop improved sensor technologies for characterization and monitoring to improve understanding and reduce uncertainties in the current nature and extent of contamination and for long-term stewardship applications.<br>• Develop improved understanding of the long-term performance of caps, in-situ containment, and stabilization technologies to reduce the likelihood of failure and re-remediation in the future, and to reduce costs associated with long-term monitoring. |
| Deactivation & Decommissioning (D&D) | **Short-Term Gaps**<br>• Robotic/remote systems and their components to improve worker safety, reduce dose, and increase productivity.<br>• Protection of workers in the field from industrial hazards<br><br>**Long-Term Gaps**<br>• Real-time characterization and decontamination of volumetrically contaminated materials to recycle and reuse materials within DOE.<br>• Cost-effective and efficient characterization and decontamination of surface-contaminated materials to free-release limits.<br>• Efficient segregation of waste materials to properly | • Reduce the costs anticipated during the D&D of surplus facilities through the deployment of improved technologies in the future.<br>• Leverage progress made in the DOE Robotics and Intelligent Machines Program.<br>• Transfer lessons learned from D&D of facilities into new practices of sustainable construction and facility operations that take into account the full life-cycle costs of construction, operation, dismantlement, and site stewardship during facilities planning. New DOE facilities and utility power plants can more adequately plan for facility decommissioning and waste management during the design phase of the new facilities. |

| | | |
|---|---|---|
| | - classify wastes and provide opportunities for recycling.<br>- Improved automation and intelligence of robotics and remote systems to improve safety, efficiency, and productivity.<br>- Integration of robotics and characterization systems to preclude the need to place workers into hostile environments.<br>- Low-cost, remote-operated size reduction and demolition equipment.<br>- Low-cost, integrated surveillance and maintenance systems. | |
| Manage Mixed Low-Level Waste (MLLW)/TRU | - Alternatives to incineration for treatment of MLLW.<br>- Long-term research that supports the economical treatment of future sources of MLLW, including, but not limited to, the large volume of disparate wastes from D&D activities | - Reduce the inventory of stored wastes.<br>- Develop alternatives to incineration for treatment of MLLW.<br>- Develop suitable handling/transportation packaging systems that enable intermodal (rail/truck) transportation options to be employed (short-term opportunity). |
| Manage High-Level Waste (HLW) | - Retrieval of HLW from difficult (single shell and obstructed) tanks, particularly those with potential leaks and those where the waste is difficult to mobilize.<br>- Treatment process for the balance of the Hanford HLW (Phase II) (e.g., advanced separations techniques for actinides and fission products).<br>- Treatment process for INEEL calcine waste.<br>- Methods to improve tank integrity and leak mitigation, including life extension methods, corrosion studies, tank integrity detection techniques, and barriers for leaks. | - Invest in strategic or longer-term R&D in separations, higher waste loading, and reduced HLW waste form volumes to achieve reductions in life-cycle costs for treatment of HLW.<br>- Develop new generation of technologies, processes, and operating regimes that could significantly reduce costs and technical risks. |

*continues on next page*

TABLE C.1 Continued

| Technical Category | Gaps | Opportunities |
|---|---|---|
| Manage High-Level Waste (HLW) | • Methods to close a tank from which wastes have been retrieved. | |
| Manage Spent Nuclear Fuel | • Lack of final waste acceptance criteria. | • Develop multi-purpose canisters and improved characterization and monitoring technologies to reduce life-cycle costs and worker exposure. |
| Manage Nuclear Materials | • Innovative technologies to address unique materials.<br>• Assay technology, gas generation research, moisture measurement, and stabilization process development related to U-233.<br>• Research on hydrogen gas generation. | • Develop innovative safeguards and security technologies to reduce long-term costs for storage and management of nuclear materials.<br>• Strengthen the portfolio through improved integration among DOE offices. |
| Dispose High-Level Waste/Spent Nuclear Fuel and Nuclear Materials | • Need for closure of remaining technical issues with Nuclear Regulatory Commission staff before submittal of license application.<br>• Long-term test data to reduce uncertainty with natural and engineered barrier performance.<br>• Finalization of disposal criteria (after licensing). | • Develop better understanding of natural and engineered environments to improve performance analyses and reduce uncertainty of the performance of the site, which could reduce life-cycle costs by reducing design conservatism. |
| Dispose of TRU, Low and Mixed Low-Level Waste and Hazardous Materials | • Research to reduce uncertainties in waste system performance.<br>• Capabilities for rapid assessment of technical and performance issues during waste disposal operations. | • Develop of alternative backfill and especially alternative ways to emplace backfill, which could result in significant cost savings.<br>• Develop techniques for emplacement of remote-handled TRU and large packages.<br>• Minimize the generation of TRU, low-level, mixed low-level, and hazardous waste during ongoing DOE operations, which could reduce life-cycle costs of waste disposal.<br>• Improve engineered systems for shallow disposal of MLLW, LLW, and hazardous waste, which could reduce long-term stewardship costs. |
| Minimize Waste | • Comprehensive inventory of wastes that are being | • Improve design and construction processes to |

131

| Generation | |
|---|---|

- generated by new and ongoing DOE activities.
- Identification of significant process waste streams (i.e., those that are created during the production of the desired end projects from EM operations and R&D projects).

include pollution prevention criteria and life-cycle impacts, which could minimize the need for shielding wastes, the production of demo-lition and tank wastes, and allow for the ability to recycle metals.

- Apply private sector methodologies to EQ op-erations.
- Apply technologies developed by other DOE offices and other agencies.

**Long-Term Stewardship**

- Information and data management and dissemination technologies to reliably maintain and ensure that there is an effective means of communicating the history of each site within a stewardship function to future generations.
- Research to address the effectiveness of social insti-tutions in maintaining and propagating the institutional systems required for long-term stewardship.
- Improved understanding of subsurface and fate and transport processes at remediation sites to assist in the development of the requirements of long-term stewardship.
- New capabilities in the area of long-term system per-formance monitoring and surveillance.
- Long-term barrier design, performance testing, and alternatives development to address the varied site configurations and environmental conditions that stewardship will have to address.
- Computer models to address risk and cost, and pre-dict system performance to allow effective long-term stewardship option selection and improve the ability to develop better total life-cycle costs.
- Cost-effective methods and technologies for mainte-nance and monitoring of stabilized sites.

ENVIRONMENTAL MANAGEMENT ADVISORY BOARD
TECHNOLOGY DEVELOPMENT AND TRANSFER COMMITTEE
## U.S. Department of Energy

October 10, 2000

Dr. David Bodde, Co-Chair          Mr. Joel Bennett, Co-Chair
EM Advisory Board                      EM Advisory Board
U.S. Department of Energy          U.S. Department of Energy
1000 Independence Ave., SW        1000 Independence Ave., SW
Washington, D.C. 20585             Washington, D.C. 20585

SUBJECT:  Review of the *"Adequacy Analysis of the Environmental Quality Research & Development Portfolio"* (September 2000)

Dear Dr. Bodde and Mr. Bennett:

This letter provides the results of a review of the subject document that was recently conducted by the Technology Development and Transfer (TD&T) Committee of the Environmental Management Advisory Board (EMAB). Mr. Gerald Boyd, Deputy Assistant Secretary, Office of Science and Technology, requested the review.

**BACKGROUND**

The *Adequacy Analysis* was prepared under the leadership of the Strategic Laboratory Council (SLC) and was released as a final report in September 2000. This SLC effort was co-chaired by Dr. Paul Kearns of the Idaho National Engineering and Environmental Laboratory (INEEL) and Dr. James Helt of Argonne National Laboratory (ANL). The stated purpose of the document was to determine the adequacy of DOE's research & development portfolio in providing the science and technology required to achieve the strategic goals and objectives of DOE's Environmental Quality (EQ) business line.

October 10, 2000
Page 2.

The document was developed with the participation of people drawn mostly from national laboratories, large EM sites, and DOE's Office of Environmental Management, Office of Science, and Office of Civilian and Radioactive Waste Management. In addition, one representative each from the Environmental Protection Agency and the Department of Defense participated, as well as several persons not affiliated with DOE.

**CHARGE TO THE TD&T COMMITTEE**

Mr. Boyd's charge to the TD&T Committee for the review involved three aspects:

1. Does the Committee think the process used in developing the document was adequate?

2. What is the Committee's opinion about the results of the analysis?

3. Finally, does the Committee have any recommendations with regard to the analysis?

**TD&T REVIEW PROCESS**

Members of the TD&T Committee met in Washington, D.C. on October 3-4, 2000. The first day of the review involved a set of interactive discussions with OST's senior management team, Drs. Kearns and Helt of the SLC, and senior technical persons representing various contractors at Hanford, Savannah River, and Idaho, who had either participated in the analysis or were knowledgeable about the results. During the meeting, we also received a progress report from Greg Symmes of the National Research Council (NRC), who is directing a related effort on EM's R&D Portfolio that is underway at NRC.

October 10, 2000
Page 3.

The Committee appreciated the participation of so many key individuals in this review and benefited greatly from the discussions that took place. Based on the information and views exchanged, Committee members were readily able to address all elements of the charge. The Committee's findings and recommendations related to each element are provided below. An agenda and committee membership list are attached.

**FINDINGS AND RECOMMENDATIONS**

## Charge 1: Adequacy of the process used to develop the analysis.

The impact of future adequacy analyses will be more far-reaching if conducted earlier in the budgetary cycle, and if more time is provided to enable a comprehensive understanding of adequacies and gaps to be developed. All participants in the review agreed that the adequacy analysis had been conducted over a relatively short timeframe. Nevertheless, the Committee found that the process used to develop the results had many positive elements, yielded a useful product that can be built upon in the future, and was generally adequate. We recognized that this was the first time an adequacy analysis of the EQ R&D Portfolio had ever been undertaken by DOE. This, in itself, represents a major step forward. The SLC (and especially Drs. Kearns and Helt) should be commended for taking the leadership on this effort and for arranging the excellent facilitating support from the INEEL, which allowed the participants to work quickly and efficiently.

It was further clear to the Committee that the interactions that had taken place among the various participants during development of the analysis was a very valuable aspect in arriving at the final results. The involvement of a cross-section of EM-savvy individuals

October 10, 2000
Page 4.

from different organizations for an EM corporate purpose proved highly beneficial and yielded additional perspectives that are usually not attained by a top-down or bottom-up analysis of this type.

The final document provides many useful insights and recommendations that can guide a stronger R&D program for EM. Overall, the Committee found that the process directed by the SLC produced a positive document that lends credibility and bolsters the rationale for many parts of the OST program.

Although the Committee believes that the results of the analysis are valuable, the Committee also thinks the process would benefit in the future by including more reviewers not directly responsible for the work being analyzed. The group of participants could be considered to lack full objectivity for the adequacy analysis since many of their organizations conduct the work that was analyzed. While the commitment of the participants to an EM corporate perspective during the analysis was evident and should be congratulated, the Committee noted that the vast majority of the participants are directly linked to DOE, so some could interpret the results as lacking certain independence.

The Committee recognizes that DOE has artificially confined the scope of the EQ business line, and therefore, this limits what the EQ R&D Portfolio can include. Obviously, this was a major constraint to conducting a comprehensive adequacy analysis of the portfolio for the first time. We take this opportunity to reiterate our previously expressed conclusion that DOE needs to broaden the definition of the EQ business line and integrate it with relevant parts of DOE's other business lines.

October 10, 2000
Page 5.

## Charge 2: Opinion on results of the adequacy analysis.

The Committee generally agreed with the results of the overall adequacy analysis, especially the fact that the R&D Portfolio has a short-term focus and lacks a longer-term strategic vision. We agree that the area of Environmental Remediation, which includes the whole myriad of major subsurface issues that remain to be understood, and the area of Managing High Level Waste are the areas that contain the most significant gaps that need to be addressed by the R&D Portfolio. We also agree that the area of Deactivation/Decommissioning supports the major EM objective of Remediating Sites and Facilities but has not yet received adequate attention from the portfolio.

The Committee found that the revised framework for the R&D Portfolio developed by the participants during the adequacy analysis was a significant improvement over the original framework and should be adopted by DOE. The three elements (Cleanup the Legacy, Disposition Wastes and Unneeded Materials, and Manage Future Risk) and five objectives (linked to individual technical categories) that were defined to support the revised framework do a much better job of communicating what the portfolio is all about. The elements also provide an excellent basis for formulating a more compelling message about the contents of the portfolio, developing a better rationale for it, and broadening support.

October 10, 2000
Page 6.

The Committee also found that defining two new technical categories for the Portfolio (Minimize Waste Generation and Long Term Stewardship) was a very positive outcome. Both of these categories highlight the evolving EQ responsibilities of DOE, especially regarding EM sites. With respect to these two categories, however, the Committee was concerned that the element under which they are found in the revised framework (i.e., Manage Future Risk) could be interpreted more like "Manage Risk in the Future." It is critical that this interpretation not be conveyed because, while both waste minimization and long-term stewardship are more focused on the future, R&D efforts on their behalf need to start now. The message should be that future programmatic risk must be managed starting now. Unfortunately, the Committee could not agree on a crisp re-wording of this element so that the wrong message was not conveyed. This may be worthy of further consideration as the Portfolio is revisited.

Additionally, the Committee is aware of efforts underway within EM (as well as within EMAB) to increase the visibility and impact of efforts involving Environment, Safety, and Occupational Health (ESOH) in the R&D Portfolio. Nevertheless, we noted that ESOH issues were still not sufficiently evident in the results of the current adequacy analysis. Given the current DOE emphasis on this topic, we believe it would be well for EM to consider how relevant ESOH issues are being addressed as part of the EQ R&D Portfolio.

The Committee also considered and discussed individually each of the seven Findings presented in the *Adequacy Analysis*. The first four Findings relate to the R&D Portfolio, while the remaining three relate to operational practices. The Committee spent most of its time considering the Findings involving the R&D Portfolio. Our comments on these four Findings are presented below. For clarity, each Finding is re-stated from the final report before our

October 10, 2000
Page 7.

comments are presented. For the record, the Committee generally concurred with the three Findings on operational practices without significant comment.

*"Finding 1: The EQ Portfolio has significant gaps and, as a whole, is underinvested."*

**Committee comments** -- While the Committee generally agreed with this Finding, we also found ourselves agreeing that a compelling case for greater investment in the Portfolio still has not been made by EM. Given the scale of the challenge facing EM, we believe that such a case can be made, even considering the lack of definition of such factors as the EM baseline, site end-states, risks, long-term budgets, political support, and appropriate contract incentives. These are realities whose existence needs to be acknowledged but which should not be used as an excuse for failing to support science and technology in EM with sound rationale and planning.

The Committee has been encouraged by the progress we have seen within EM during the past few years regarding science and technology and the new mechanisms that are being put into place. These include the development of roadmaps, development of waste disposition maps, increased use of projectization, and R&D Portfolio planning and analysis. The Committee believes the supporting case for increased R&D investment needs to be made in terms of real payoff to the country. In this context, participants in the EQ R&D Portfolio need to clearly move away from a community entitlement mentality as the basis for receiving increased investment. This means moving from thinking like *"We should receive 'X' percent of the overall budget for R&D purposes."* to a value-added approach that emphasizes something like *"Our R&D efforts will address and resolve these critical public and environmental health, cost, and schedule risks."*

October 10, 2000
Page 8.

*"Finding 2: The R&D portfolio does not include a longer-term vision and 'strategic' elements such as alternative technologies and next-generation solutions."*

<u>Committee comments</u> -- The Committee agreed with this Finding and believes it is not only a manifestation of the under-investment problem but also of the cultural and financial situation in which EM finds itself, governed by compliance agreements that were formulated independently of current budgetary and technical realities.

Further, the Committee believes that science and technology (S&T) continues to be under appreciated within EM as the source of needed long-term solutions. While this situation has clearly improved during the tenure of Undersecretary Moniz, we are concerned that some of the positive recent impacts and advances we have seen may not become more solidly institutionalized.

*"Finding 3: The funding distribution across the technology maturity spectrum is unbalanced."*

<u>Committee comments</u> -- The bimodal funding distribution, in which DOE's investments in S&T are focused on basic research and demonstration/deployment activities, leaves a gap in applied research and development. The Committee believes that this is another manifestation of under-investment. However, it also reflects EM's reaction to the pressure from Congress to show more deployments (i.e., more payoff from past investments). Further, it indicates that EM has still not developed an integrated S&T program that links basic and applied research seamlessly with development and deployment efforts that address and solve problems in the field.

October 10, 2000
Page 9.

The Committee is convinced that the imbalance in funding distribution cannot be successfully addressed unless "users" are more effectively involved in the overall S&T process from the beginning. Users in EM have consistently demonstrated that they are willing to co-invest with OST in such programs as the Technology Deployment Initiative (TDI) and Accelerated Site Technology Deployment (ASTD). However, these programs have still not become firmly institutionalized. In addition, DOE has not fully supported adequate funding from Congress for the EM Science Program and has seen funding for this program decline steadily. The current increase in the FY01 budget for OST proposed by Congress is heartening to the Committee. Hopefully, this will provide EM with a further opportunity to move toward a more coherent, integrated, seamless, effective S&T program.

*"Finding 4: Significant life-cycle costs and corresponding R&D hinge on highly uncertain end-states."*

<u>Committee comments</u> -- This Finding appears to be a fact-of-life in the EM world that must be accepted and continually dealt with. Rather than dealing with the often-elusive concept of defining "end-states," which are often decades away, it may be more useful to focus on defining a series of more limited "intermediate-points" or "end-points," the sum total of which can eventually lead to an end-state. We believe that end-points can potentially be better defined, and they lend themselves to better overall management and measurement of progress. More precise terminology may also build more credibility with Congress and assist in making a case for more funding for technology needs.

<u>**Charge 3: Recommendations about the adequacy analysis.**</u>

The Committee's recommendations regarding the adequacy analysis are presented below.

October 10, 2000
Page 10.

1. DOE should institutionalize the process of conducting an adequacy analysis of the EQ R&D Portfolio. This effort should become a deliberate and formal process, and adequate time and resources should be allocated for it.

2. EM (OST) should accept the results of the first adequacy analysis and use them in a proactive way to improve its R&D Portfolio.

3. EM should perform an adequacy analysis of its R&D Portfolio at least every two years.

4. The community of participants used to develop an adequacy analysis should be broadened to enhance the credibility and perspective (objectivity) of the Portfolio and the results. The participants should include a limited number of external independent experts.

5. EM still needs to focus on more effective ways to define and support the expected payoff from the OST program. The waste disposition roadmaps developed by the INEEL should be more widely used as the basis for helping to define where OST should be making its S&T investments.

This concludes our comments and recommendations. The Committee very much appreciated the opportunity to conduct this review and offer our views for consideration by EM. We received excellent cooperation from OST management, as well as from the SLC and senior individuals from the DOE contractor community.

October 10, 2000
Page 11.

We are encouraged by the attention being given to improving the S&T program and look forward to working with EM on the whole range of issues represented by the EQ R&D Portfolio.

Sincerely yours,


Edgar Berkey, Ph.D.
Chairman
Technology Development & Transfer Committee

cc:     James Melillo, DOE-EM, EMAB
        TD&T Committee Members

Attachments     [not included in appendix]

# APPENDIX D

# DESCRIPTIONS OF DOE'S ENVIRONMENTAL QUALITY TECHNICAL CATEGORIES

These descriptions are based largely on those in the Department of Energy's (DOE's) Environmental Quality (EQ) research and development (R&D) portfolio document (DOE, 2000b) and are intended to provide the reader with an overview of the magnitude and duration of DOE's "EQ challenges" (see Sidebar 2.3). They are not intended to represent a comprehensive description of the problem areas or the types of R&D activities currently being conducted by DOE.

## Manage High-Level Waste

High-level waste (HLW) is highly radioactive material resulting from reprocessing of spent nuclear fuel, which includes both liquid waste and solid residues. Large quantities of HLW were generated during production of nuclear weapons and reprocessing of defense production reactor fuels. There are 280 large radioactive waste storage tanks and more than 63 smaller underground storage tanks across the DOE complex that contain more than 340,000 cubic meters (90 million gallons) of HLW waste. Most of these tanks have exceeded their design life, some have leaked, and all represent potential occupational and public risks.

The waste is currently stored at five main locations in both solid and liquid form: (1) Savannah River, South Carolina; (2) Hanford, Washington; (3) Idaho National Engineering and Environmental Laboratory (INEEL); (4) Oak Ridge Reservation, Tennessee; and (5) West Valley Demonstration Project, New York. To protect the public and the environment, much of this waste must be retrieved from the tanks and converted into an appropriate form for long-term disposal. Some HLW has been immobilized in glass at Savannah River and West Valley. DOE has signed federal facility agreements with state and federal regulators that drive the scope and schedule for cleanup and closure of the tanks. DOE estimates that HLW cleanup will continue until at least 2046, at a total projected life-cycle cost of $54 billion. In fiscal year 2000, DOE spent approximately $57.6 million on R&D to address needs related to the management of high-level waste. DOE also recognizes that after cleanup most sites that stored HLW will require long-term institutional management measures indefinitely to protect human health and the environment (see "Long-Term Institutional Management" below).

## Manage Mixed Low-level/Transuranic Waste

Mixed low-level waste (MLLW) is low-level waste that contains both chemically hazardous and radioactive components. Transuranic (TRU) waste is any waste, except for HLW, containing more than 100 nanocuries per gram of long-lived (>20 years), alpha-emitting TRU radionuclides. TRU waste is produced primarily from reprocessing of irradiated fuel and fabrication of nuclear weapons and contains isotopes such as plutonium and americium. Unlike HLW, TRU waste is non-heat bearing. Low-level waste is waste that is not spent fuel, HLW, or uranium or thorium mill tailings.

Thirty-six DOE sites store about 165,000 $m^3$ of mixed low-level and transuranic waste. Considerable amounts of TRU waste also contain hazardous constituents subject to regulation under the Resources Conservation and Recovery Act (RCRA) or the Toxic Substances Control Act. Since 1970, DOE has placed TRU waste in retrievable storage, such as metal drums or boxes, either on storage pads, in buildings, or in tanks. TRU waste is managed at 21 sites. DOE has begun disposal of stored post-1970 TRU waste at the Waste Isolation Pilot Plant (WIPP) near Carlsbad, New Mexico. Because MLLW contains chemically hazardous as well as non-transuranic radioactive materials, it is subject to regulation under both RCRA and the Atomic Energy Act. The storage, treatment, and disposal of MLLW are subject to state and federal regulations. The estimated life-cycle cost for management and disposition of mixed low-level and TRU waste is more than $18 billion. In fiscal year 2000, DOE spent approximately $29.1 million on R&D related to the management of mixed low-level/TRU waste.

## Manage Spent Nuclear Fuel

Spent nuclear fuel (SNF) is irradiated nuclear fuel that has not been reprocessed. The United States operated 14 nuclear defense production reactors between 1944 and 1988 to produce plutonium and tritium for nuclear warheads. In addition, the United States operated many other test reactors to encourage and support both commercial and military reactor developments. (The spent nuclear fuel arising from the operation of commercial nuclear power plants is described below.) During that time, most of the nuclear fuel rods and targets irradiated in the reactors were reprocessed to extract the plutonium or tritium and the remaining enriched uranium for reuse. In addition, the U.S. Navy operated many nuclear propulsion reactors from which the fuel assemblies were processed to recover and reuse the remaining fissile uranium. DOE's SNF is not categorized as waste, but it is highly radioactive and must be stored in special facilities that shield and cool the material. Most SNF is stored in indoor pools under water, although some spent fuel is kept in dry storage.

Three DOE sites (INEEL, Savannah River, and Hanford) manage most of the SNF in the DOE complex. Hanford has an inventory of over 2,100 metric tons heavy metal (MTHM) of SNF from its production reactors. After washing, packaging, and drying, this SNF will be transferred to dry storage until shipment (either to a repository or to an alternative treatment system). INEEL has an inventory of 270 MTHM of SNF, and expects to receive an additional 60 MTHM. After on-site storage, drying, and packaging, all SNF is expected to be shipped off-site to a repository for disposal. Savannah River has an inventory of 20 MTHM, and expects to receive an additional 30 MTHM from off-site sources. The SNF is expected to be prepared and placed in an off-site geologic repository (the same one as for commercial spent fuel and HLW). The total life-cycle cost for management and preparation for disposal of DOE's SNF is estimated to be about $7 billion (DOE, 2000b). In fiscal year 2000, DOE spent approximately $12 million on R&D related to the management of spent nuclear fuel.

## Manage Nuclear Materials

A major consequence of the end of the Cold War has been a decrease in the number of U.S. nuclear weapons deployed around the world. This decrease resulted in nuclear weapons components being returned to DOE and classified as surplus materials (approximately 200 metric tons of U.S. weapons-usable fissile materials, which includes highly enriched uranium and plutonium, are classified as surplus materials). Disposition of this surplus material will be carried out either by making it into reactor fuel and burning it in electricity-producing commercial reactors (producing spent fuel) or by immobilizing the material mixed with high-level waste. In both cases, the resulting materials will be prepared for disposal in the geological repository.

Other nuclear materials are present in weapons complex facilities that were shut down in the late 1980s and early 1990s due to concerns over safety and environmental problems, and the end of the Cold War.

DOE also has an inventory of over 700,000 metric tons of depleted uranium hexafluoride and a variety of special purpose isotopes like U-233. The estimated life-cycle cost for management and disposition of DOE's nuclear materials is approximately $7 billion (DOE, 2000b). In fiscal year 2000, DOE spent approximately $7.6 million on R&D related to the management of nuclear materials.

## Dispose of High-Level Radioactive Wastes, Spent Nuclear Fuels, and Nuclear Materials

DOE is responsible for providing for the permanent disposal of U.S. high-level radioactive waste and SNF (Public Law 97-425). The Yucca Mountain Site in Nevada has been designated as the only site to be characterized to determine its suitability for a geologic repository (Public Law 100-203). The types of waste that will be disposed of in the geologic repository consist of commercial spent fuel (including mixed oxide spent fuel [i.e., fuel that contains both uranium and plutonium from weapons dismantlement]), high-level waste (including immobilized plutonium), and DOE spent fuel (including naval spent fuel). Other wastes, such as greater-than-class-C, may also be disposed of in the repository.

Commercial spent fuel consists of fuel assemblies discharged from electricity-generating nuclear reactors and is located at 72 nuclear power plant sites and one independent storage site in 33 states. The total inventory of spent fuel at the end of 1998 was estimated to be about 38,000 MTHM, and the expected inventory in 2040 is projected to be about 85,000 MTHM. High-level waste to be disposed of is immobilized (generally as a borosilicate glass or a ceramic) and encased in metal canisters. It is estimated that approximately 22,000 canisters will be produced through 2035 (including those that will contain immobilized surplus weapons-usable plutonium). The DOE spent fuel inventory projected to the year 2035 is estimated to be 2,500 MTHM.

DOE plans to submit a site suitability recommendation for the Yucca Mountain Site to the President in 2001, and if the site is determined to be suitable and approved by both the President and Congress (after presidential approval, the state of Nevada can submit a notice of disapproval that can be overridden by a majority vote of both houses of Congress), to prepare and submit a license application to the U.S. Nuclear Regulatory Commission in 2003 for construction authorization for the repository. To obtain the license, DOE must demonstrate that a repository can be constructed, operated, monitored, and eventually closed without unreasonable risk to the health and safety of workers and the public. The repository schedule calls for initial waste emplacement in 2010, followed by several decades of operation and further decades of monitoring and performance confirmation. In fiscal year 2000, DOE spent approximately $47 million on R&D to address needs related to the disposal of high-level radioactive waste, spent nuclear fuels, and nuclear materials.

## Environmental Remediation of Contaminated Sites (Lands and Waters)

Environmental remediation involves the removal or stabilization of radioactive and/or hazardous contaminants in soil, fractured bedrock, and groundwater. The primary objectives are to identify, contain, remediate, and remove contamination, and to validate that environmental remediation has achieved the desired end state. Approximately 3 million cubic meters (100 million cubic feet) of solid radioactive and hazardous wastes are buried in the subsurface throughout the DOE complex. The largest contamination challenges are at the INEEL, Oak Ridge, Hanford, Rocky Flats, and Savannah River sites. Contaminants are located in the subsurface both above and below the water table. DOE estimates that 75 million cubic meters (2.6 billion cubic feet) of soil and 1.8 billion cubic meters (475 billion gallons) of groundwater are contaminated and require remediation. Contaminants include hazardous metals such as chromium, mercury, and lead; radioactive laboratory and processing waste; explosive and pyrophoric materials; solvents; and numerous radionuclides. The total life-cycle cost of environmental remediation activities through 2070 is estimated to be greater than $13 billion (DOE, 2000b). In fiscal year 2000, DOE spent approximately $52 million on R&D related environmental remediation of contaminated DOE sites.

## Deactivation and Decommissioning of Contaminated Facilities

Many of the more than 20,000 DOE facilities that were used to support nuclear weapons production and other activities are contaminated with radioactive materials, hazardous chemicals, asbestos, and lead. To reduce the potential for release of radioactive and hazardous materials to the environment, the risk of industrial safety accidents, and the costs of monitoring and maintaining these facilities, DOE plans to deactivate and decommission (D&D) such facilities. Deactivation is defined as activities to reduce the physical risks and hazards at these facilities, to reduce the costs associated with monitoring and maintenance of these facilities (i.e., facility mortgage), and make these facilities available for potential reuse or eventual decommissioning. Decommissioning is defined as activities associated with decontamination, demolition, and final disposition of the facility and the equipment contained within. The estimated life-cycle cost of D&D

activities for facilities currently under DOE responsibility is $12.5 billion. In fiscal year 2000, DOE spent approximately $12.7 million on R&D to address needs related to the deactivation and decommissioning of contaminated DOE facilities.

## Long-Term Stewardship

Of the 144 contaminated sites currently under its control, DOE estimates that fewer than 25 percent will be cleaned up sufficiently to allow unrestricted use. At many sites, radiological and non-radiological hazardous wastes will remain, posing risks to humans and the environment for tens or even hundreds of thousands of years. For these sites, a broad-based, systematic approach that integrates contaminant reduction, contaminant isolation, and stewardship will be required to protect human health and the environment (NRC, 2000a; DOE, 1999a, 2001b). DOE estimates that it currently spends approximately $64 million annually on long-term stewardship activities, and these costs will increase to nearly $100 million annually by 2050, when all sites are expected to be closed (DOE, 2001b).

## Minimization of the Risk of Newly Generated Radioactive and Hazardous Waste

The recent adequacy analysis of the EQ R&D portfolio (DOE, 2000g) recommended that a new category of R&D activities be defined to minimize the risk of newly generated DOE radioactive and hazardous waste. DOE currently has no complex-wide R&D program to minimize the generation of new wastes, although site specific work is in progress to address local waste management programs (DOE, 2000g).

# APPENDIX E

# DESCRIPTIONS OF RELATED RESEARCH AND DEVELOPMENT PROGRAMS

As part of its information-gathering activities, the committee considered research and development (R&D) programs in other federal agencies, such as the Department of Defense (DOD) and the Environmental Protection Agency (EPA). The committee also considered a number of relevant international R&D programs. Although the committee did not conduct a comprehensive examination of national and international R&D programs, it did identify a number of programs that support R&D relevant to the Department of Energy's (DOE's) Environmental Quality (EQ) mission.

## U.S. DEPARTMENT OF DEFENSE

The **Strategic Environmental Research and Development Program** is DOD's environmental R&D program, operated jointly with DOE and EPA, with participation by numerous other federal organizations. The program focuses on cleanup, compliance, conservation, and pollution prevention technologies. The development and application of innovative environmental technologies is intended to reduce costs, environmental risks, and/or the time required to resolve environmental problems in these areas while enhancing safety and health. Equally important, the development and application of innovative pollution prevention technologies serves to reduce or eliminate waste problems before they occur. Examples of research emphases are the areas of site characterization and monitoring, remediation, and risk assessment. The total fiscal year 2001 budget is $59.6 million.

The **Environmental Security Technology Certification Program** demonstrates and validates promising, innovative technologies that target DOD's most urgent environmental needs. These technologies are intended to provide a return on investment through cost savings and improved efficiency. Projects are selected in the areas of cleanup, compliance, pollution prevention, and detection and remediation of unexploded ordinances. Technologies are demonstrated and evaluated at DOD sites and effective and affordable technologies are transferred across DOD.

The **Defense Advanced Research Projects Agency,** the central R&D organization for DOD, manages and directs basic and applied R&D projects, and pursues research and technology where risk and payoff are both high and where success may provide advances for traditional military roles and missions. Its mission is to develop imaginative, innovative, and often high-risk research ideas offering a significant technological impact that will go well beyond the normal evolutionary developmental approaches and to pursue these ideas from the demonstration of technical feasibility through the development of prototype systems.

The **Toxic Biological Interactions program** of the U.S. Air Force Office of Scientific Research supports basic research that endeavors to understand how such toxic agents as heavy metals (chromium and cadmium) and various chemicals that constitute fuels, propellants, and lubricants may interact with biological systems at the subcellular and molecular levels to produce toxic effects. The Air Force also supports studies that explore novel experimental and computational techniques for assessing the potential health risks of these agents.

The **Surface and Interfacial Chemistry Program** of the Army Research Office supports research on the decomposition of hazardous molecules on well-characterized surfaces and in organized media (e.g., micelles, microemulsions, vesicles, and monolayer films) at liquid-liquid and liquid-solid interfaces. The development of new experimental probes of these reactions is also of interest. The most important species are organo-phosphorus, -sulfur, and -nitrogen molecules and reactions of organic functional groups on surfaces and in these organized media. The principle reactions of interest are hydrolysis and oxidation, and catalysis is a strongly desired goal of these studies; however, new concepts are encouraged.

The **Mechanical and Environmental Sciences Division** of the Army Research Office supports basic research related to the remediation and restoration of sites contaminated by Army actions and the use of military training lands. The Army Research Office also supports the Research and Technology Integration Directorate, which integrates scientific research and technology.

The Office of Naval Research sponsors an **Environmental Quality Program** that is aimed at developing technology leading to affordable environmental compliance and pollution prevention. The program supports basic research, applied research, and advanced technology development. Program areas include environmental chemistry (basic research), applied research, and environmental requirements advanced technology. The program focuses on technologies directed toward environmentally sound ships, shore-related facilities, and sediment issues, and specific research interests include sensors and improved cleaning methods.

## U.S. ENVIRONMENTAL PROTECTION AGENCY

EPA's R&D is funded primarily through its Office of Research and Development (ORD). ORD conducts leading-edge research and fosters the use of science and technology in fulfilling EPA's mission to protect human health and safeguard the environment. It operates several research laboratories across the country that specialize in specific areas of R&D.

The **National Exposure Research Laboratory**, conducts R&D that leads to improved methods, measurements, and models to assess and predict exposures of humans and ecosystems to harmful pollutants and other conditions in air, water, soil, and food.

The **National Risk Management Research Laboratory** conducts research into ways to prevent and reduce risks from pollution that threaten human health and the environment. The laboratory investigates methods and their cost-effectiveness for prevention and control of pollution to air, land, water, and subsurface resources; protection of water quality in public water systems; remediation of contaminated sites, sediments and groundwater; prevention and control of indoor air pollution; and restoration of ecosystems. The goal of this research is to provide solutions to environmental problems by developing and promoting effective environmental technologies; developing scientific and engineering information to support regulatory and policy decisions; and providing the technical support and information transfer to ensure implementation of environmental regulations and strategies at the national and community levels.

The **Superfund Innovative Technology Evaluation Program** was established by EPA's Office of Solid Waste and Emergency Response and ORD in response to the 1986 Superfund Amendments and Reauthorization Act, which recognized a need for an alternative or innovative treatment technology research and demonstration program. The program is administered by ORD's National Risk Management Research Laboratory.

The National Center for Environmental Research sponsors environmental research grants under the **Science to Achieve Results Program**. Included are fellowships for graduate environmental study and minority academic institutions fellowships for graduate environmental study.

The **Environmental Technology Verification Program** was instituted to verify the performance of innovative technical solutions to problems that threaten human health or the environment. The program was created to substantially accelerate the entrance of new environmental technologies into the domestic and international marketplace. It verifies commercial-ready, private sector technologies through 12 pilots.

The **Subsurface Protection and Remediation Division** of the National Risk Management Research Laboratory conducts research and engages in technical assistance and technology transfer on the chemical, physical and biological structure and processes of the subsurface environment, the biogeochemical interactions in that environment, and fluxes to other environmental media.

The **Waste Research Strategy** covers research necessary to support both the proper management of solid and hazardous wastes and the effective remediation of contaminated waste sites. This research includes methods to improve the assessment of existing environmental risks and to develop more cost-effective ways to reduce those risks. This strategy focuses on the following research areas: contaminated groundwater, contaminated soils and the vadose zone, emissions from waste combustion facilities, and active waste management facilities.

The **National Center for Clean Industrial and Treatment Technologies** is a research consortium dedicated to advancing science, engineering, and pollution prevention, established through a base grant from EPA's Centers Program. Since its establishment, the center has initiated 57 projects involving 51 principal investigators, 57 companies, 33 government and other organizations, and well over 100 students. Targeted industry sectors have included chemical processing, metals, manufacturing, energy, and forest products. Participating disciplines have included environmental, chemical, civil, mechanical, metallurgical and geological engineering; chemistry; biology; social science; business; and forestry.

One of the programs sponsored by EPA's National Center for Environmental Research and Quality Assurance is the **Hazardous Substance Research Centers Program**. The mission of the program is to conduct research to develop and demonstrate new methods to assess and remediate sites contaminated with hazardous substances, improve existing treatment technologies, decrease the production and use of hazardous substances, educate hazardous substance management professionals, and improve community public awareness. The program provides basic and applied research, technology transfer, and training and encourages integrated research projects. The program consists of five multi-university centers, which are located in different regions and focus on different aspects of hazardous substance management. EPA, DOE, DOD, academia, and other federal agencies fund the centers. A description of these centers is found in Sidebar E.1.

DOE's Office of Science and Technology and EPA's Office of Solid Waste recently signed a memorandum of understanding (MOU) to improve cooperation on the development of technical solutions to problems associated with mixed wastes. The main objective of the MOU is to provide the Office of Solid Waste with performance and cost data from the demonstration and field testing of mixed waste treatment and control technologies, which is expected to help EPA develop sound and cost-effective regulations and standards for mixed wastes. The effort also is intended to facilitate cooperation in budgetary planning for OST's R&D efforts and EPA's regulatory activities.

## U.S. NUCLEAR REGULATORY COMMISSION

The U.S. Nuclear Regulatory Commission's Radiation Protection, Environmental Risk and Waste Management Branch develops, plans, and manages research programs related to the movement of radionuclides in the environment and consequent dose and health effects to the public and workers as a result of nuclear power plant operation, facility decommissioning, cleanup of contaminated sites, and disposal of radioactive waste.

## U.S. GEOLOGICAL SURVEY

The **Toxic Substances Hydrology Program** provides scientific information needed to improve characterization and management of contaminated sites, to protect human and environmental health, and to reduce potential future contamination problems. The goal of the program is to provide scientific information on the behavior of toxic substances in

---

**SIDEBAR E.1 EPA's Hazardous Substance Research Centers**

The **Great Lakes and Mid-Atlantic Center** focuses on remediation of hazardous organic compounds found in soil and groundwater. Ongoing research focuses on in situ bioremediation, surfactant introduction, and bioventing technologies. The lead institution is the University of Michigan, and other participating universities include Howard University and Michigan State University.

The **Great Plains/Rocky Mountain Center** focuses on contaminated soils and mining wastes. Research covers soil, water, and groundwater contaminated with heavy metals, and organics; wood preservatives in groundwater; pesticides; improved methods for analyzing contaminated soil; and pollution prevention technologies. The lead institution is Kansas State University, and other participating universities include Haskell Indian Nations University, Kansas State University, Lincoln University, and Montana State University.

The **Northeast Center** focuses on incineration/thermal treatment, characterization and monitoring, in situ remediation, and ex situ treatment of industrial wastes. The center is a consortium of 7 universities: the New Jersey Institute of technology (which serves as the lead institution), the Massachusetts Institute of Technology, Princeton, Rutgers, Stevens Institute of Technology; Tufts; and the University of Medicine and Dentistry of New Jersey.

The **South and Southwest Center** focuses on contaminated sediments, in particular, in situ chemical mobilization in beds and confined disposal facilities, in situ remediation, and in situ detection covers. The lead institution is Louisiana State University, and other participating universities include Georgia Institute of Technology and Rice University.

The **Western Region Center** focuses on groundwater cleanup and site remediation with a strong emphasis on biological approaches. Projects address chlorinated solvents; halogenated aromatics (pentachlorophenol and PCBs), nonhalogenated aromatics, including petroleum derivatives; ordinance wastes, heavy metals; and transport and fate. The lead institution is Stanford University, with Oregon State University participating.

---

hydrologic environments, including surface water, groundwater, soil, sediment, and the atmosphere.

The **National Water-Quality Assessment Program** is designed to describe the status and trends in the quality of ground- and surface-water resources and to provide a sound understanding of the natural and human factors that affect the quality of these resources. Regional and national syntheses of information provide summaries on volatile organic compounds, trace elements, and surface water-quality monitoring.

The **Ground-Water Resources Program** encompasses regional studies of groundwater systems, multidisciplinary studies of critical groundwater issues, access to groundwater data, and research and methods development. The program provides unbiased scientific information and many of the tools that are used by federal, state, and local management and regulatory agencies to make important decisions about groundwater resources.

The **Biomonitoring of Environmental Status and Trends** evaluates environmental contaminants and their effects on species and lands under the stewardship of the Department of Interior to provide scientific information and guide management actions. The program is designed to identify and understand the effects of environmental contaminants on biological resources, particularly those resources under the stewardship of the Department of the Interior. The program provides sound scientific information to be used proactively to prevent or limit contaminant-related effects on biological resources. The primary goals are to (1) determine the status and trends of environmental contaminants and their effects on biological resources; (2) identify, assess, and predict the effects of contaminants on ecosystems and biological populations; and (3) provide summary information in a timely manner to managers and the public for guiding conservation efforts. To address these goals, the program will use different approaches, involving a combination of field biomonitoring methods and information assessment tools, for examining contaminant issues at the national, regional, and local levels.

In addition, the U.S. Geological Survey has MOUs with a number of federal agencies. For example, an MOU with the U.S. Nuclear Regulatory Commission explored R&D in the earth sciences related to the management, disposal, and environmental remediation of nuclear and mixed wastes; site decommissioning reviews; uranium in situ mining; and uranium mill tailings at existing and future sites in the United States. An MOU with EPA addressed activities related to the protection of groundwater quality.

## NATIONAL SCIENCE FOUNDATION

The **Division of Environmental Biology** supports fundamental research on the origins, functions, relationships, interactions, and evolutionary history of populations, species, communities, and ecosystems. The division also supports a network of long-term ecological research sites, doctoral dissertation research, and research conferences and workshops. Funding for fiscal year 2000 was $89.8 million.

Basic research in the **Directorate for Geosciences** advances scientific knowledge of Earth's environment, including resources such as water, energy, minerals, and biological diversity. The funding level for earth sciences was $102 million for fiscal year 2000. The directorate also supports the **Biocomplexity in the Environment Program,** a set of coordinated activities in environmental science, engineering, and education, which advance scientific knowledge about the connection between the living and non-living Earth system. The Directorate of Geosciences will provide $39.50 million in fiscal year 2001 for focused biocomplexity studies, which will enable the initiation and/or enhancement of several interdisciplinary activities.

The **Environmental Engineering Program** in the Division of Bioengineering and Environmental Systems supports sustainable development research with the goal of applying engineering principles to reduce adverse effects of solid, liquid, and gaseous discharges into land, fresh and ocean waters, and air that result from human activity and impair the value of those resources. This program also supports research on innovative biological, chemical, and physical processes used alone or as components or engineered systems to restore the usefulness of polluted land, water, and air resources. Research may be directed toward improving the cost-effectiveness of pollution avoidance and developing fresh principles for pollution avoidance technologies.

The **Division of Chemical and Transport Systems** supports research that involves the development of fundamental engineering principles, process control and optimization strategies, mathematical models, and experimental techniques, with an emphasis on projects that have the potential for innovation and broad application in such areas as the environment, materials, and chemical processing. Special emphasis is on environmentally benign chemical and material processing. Research support is available in through the following activities: chemical reaction processes; interfacial, transport, and separation processes; fluid and particle processes; and thermal systems. Funding for fiscal year 2000 was $44.3 million.

The **Division of Civil and Mechanical Systems** funds research that contributes to the knowledge base and intellectual growth in the areas of infrastructure construction and management, geotechnology, structures, dynamics and control, mechanics, and materials; sensing for civil and mechanical systems; and the reduction of risks induced by earthquakes and other natural and technological hazards. The division encourages cross-disciplinary partnerships. These partnerships promote discoveries using technologies such as autoadaptive systems, nanotechnology, and simulation to enable revolutionary advances in civil and mechanical systems. Funding for fiscal year 2000 was $48.2 million.

The **Inorganic, Bioinorganic, and Organometallic Chemistry Program** in the Chemistry Division supports research on synthesis, structure, and reaction mechanisms of molecules containing metals, metalloids, and nonmetals encompassing the entire periodic table of the elements. Included are studies of stoichiometric and homogeneous catalytic chemical reaction; bioinorganic and organometallic reagents and reaction; and the synthesis of new inorganic substances with predictable chemical, physical, and biological properties. Such research provides the basis for understanding the function of metal ions in biological systems, for understanding the synthesis of new inorganic materials and new industrial catalysts, and for systematic understanding of the chemistry of most of the elements in the environment.

The **Organic Chemical Dynamics Program** also in the Chemistry Division supports research on the structures and reaction dynamics of carbon-based molecules, metallo-organic systems, and organized molecular assemblies. Research includes studies of reactivity, reaction mechanisms, and reactive intermediates, and characterization and investigation of new organic materials. Such research provides the basis for understanding and modeling biological processes and for developing new or improved theories relating chemical structures and properties. Funding for the Chemistry Division was $139 million for fiscal year 2000.

## NATIONAL INSTITUTES OF HEALTH

The **Superfund Basic Research Program** is focused on acquiring new scientific and engineering knowledge that advances both society's understanding of the human and ecological risks from hazardous substances and the development of new environmental technologies for the cleanup of Superfund sites. The knowledge acquired in this program not only serves as the basis for subsequent basic or applied research in these areas but also provides a foundation for such practical benefits as lower cleanup costs on hazardous waste sites and improvements in human and ecological health risk assessment. The program, created and administered by the National Institute of Environmental Health Sciences, receives funding from EPA through an interagency agreement using Superfund trust monies. The research efforts undertaken by this program complement activities in EPA and the Agency for Toxic Substances and Disease Registry.

## NON-FEDERAL U.S. R&D

The Electric Power Research Institute **Decommissioning Technology Program** assists utilities to minimize the cost of decommissioning through enhanced planning, determining optimum financial fund set-aside, applying lessons learned by other utilities with retired plants, and use of advanced technology. For decommissioned power plants, site characterization and final site survey have also been costly elements of their decommissioning activities. Several technical areas have been identified where improved technology could be of considerable benefit to utilities with shutdown plants by reducing labor costs, personnel exposures, and radioactive waste. Chemical decontamination developments are discussed below. Other topics under study include site characterization, fuel pool cleanup, concrete decontamination and other remediation techniques. In conjunction with the Federal Energy Technology Center, evaluation of the applicability to U.S. power plants of technology developed in DOE programs and those in other countries is being carried out, including status reports on appropriate techniques. The Strategic Science and Technology Program addresses priority needs and opportunities by integrating scientific developments and emerging technologies with strategic industry issues and the public good.

## INTERNATIONAL R&D PROGRAMS

The committee also considered a number of international programs that support R&D related to DOE's EQ mission. They cover a wide range of issues, such as chemical processes, soil remediation, hydrology, and migration of radionuclides. Some of these programs are described below.

The **Belgian Nuclear Research Centre** is a federal organization for scientific research in the field of safe and peaceful applications of nuclear energy for industrial and medical use.

**Atomic Energy of Canada Limited** is a leading vendor of nuclear power reactors, engages in a wide range of R&D activities, and provides nuclear engineering products and services worldwide to customers in nuclear and related industries.

The **National Cooperative for the Disposal of Radioactive Waste (Nagra)** in Switzerland provides the technical and scientific basis for safe management of radioactive waste. Nagra has a number of cooperative agreements with other countries, including the United States.

The **Paul Scherrer Institut** in Switzerland is the federal institute for reactor and nuclear R&D. It covers the areas of incineration of wastes; modeling of radionuclide migration through heterogeneous geologic media; chemical behavior of radionuclides during migration; transport of radionuclides through the biosphere; natural analogue studies; hydrological studies; sorption constants on different rocks; immobilization of low-level waste and intermediate-level waste in cement; leaching rates on low-level and intermediate-level waste forms; and long-term corrosion tests on waste-packaging materials.

**Nirex**, in the United Kingdom, examines safety, environmental, and economic aspects of deep geological disposal. It deals with intermediate-level waste, which accounts for the majority of radioactive waste currently in storage, and with some low-level waste.

The Canadian National Research Council's **Institute for Chemical Process and Environmental Technology** funds research in the following areas: environmental management; chemical sensors; soil remediation, computational fluid dynamics and reactive flow modeling; and chemical process simulation, design, and economics. Chemical process simulation techniques are being investigated as tools for improving process design and developing clean technology for pollution prevention and waste reduction.

The **Geological Survey of Canada** funds research in environmental geology, such as the distribution and concentration of heavy metals near mines, in its Terrain Sciences Division.

The **Environmental Agency of England and Wales** sponsors research in several areas, including waste management. Research projects cover such topics as the effects of substances in groundwater on the migration of radionuclides, national recovery and recycling database for waste management, life-cycle cost of waste management options, radionuclide migration processes in geological media, and environmental impact of old landfills.

# APPENDIX F

# ANNOTATED BIBLIOGRAPHY OF SELECTED RECENT NATIONAL RESEARCH COUNCIL REPORTS

**NUCLEAR WASTES: TECHNOLOGIES FOR SEPARATIONS AND TRANSMUTATION (NRC, 1995)**
This report describes the state of the art in separations and transmutation technologies, and considers their application to U.S. high-level radioactive waste and spent nuclear fuel. It concludes that a modestly funded research and development (R&D) program in particular technical areas is of value, but that R&D work is not sufficiently viable to justify delays in geological repository development at Yucca Mountain.

**THE WASTE ISOLATION PILOT PLANT: A POTENTIAL SOLUTION FOR THE DISPOSAL OF TRANSURANIC WASTE (NRC, 1996a)**
This report addresses the suitability of the Waste Isolation Pilot Plant (WIPP) as a geological repository for transuranic waste by examining scenarios for the possible release of radionuclides to the environment after the repository is filled and sealed. The committee's conclusions were that (1) human exposure to radionuclide releases from WIPP is likely to be low compared to U.S. and international standards and (2) if the repository were sealed effectively and undisturbed by human activity, there would be no credible or probable scenarios for release of radionuclides to the environment. The committee also made several recommendations for additional work that should be done by the Department of Energy (DOE) and its contractors to assess the likelihood of future human disturbance to the repository and to reduce the impacts of such disturbances if they occur. This report (and earlier reports by the same committee) was instrumental in DOE's efforts to gain regulatory approval to open the first U.S. geological repository. The Environmental Protection Agency also used the report in its review of DOE's license application. The WIPP repository received its first shipment of waste in early 1999.

## THE HANFORD TANKS: ENVIRONMENTAL IMPACTS AND POLICY CHOICES (NRC, 1996b)

This report reviews a draft environmental impact statement for the remediation of high-level radioactive waste in tanks at the Hanford Site, Washington. The report recommends that remediation activities use a phased decision strategy, proceeding with current cleanup operations while filling in important information gaps before making a final decision as to which technologies and methodologies will ultimately be implemented. Remediation of the tanks should be consistent with plans for the entire Hanford Site, including the environment and future land use.

## BUILDING AN EFFECTIVE ENVIRONMENTAL MANAGEMENT SCIENCE PROGRAM: FINAL ASSESSMENT (NRC, 1997)

This report summarizes the potential value of basic research to DOE's cleanup mission and advises DOE on the structure and management of its Environmental Management Science Program (EMSP). The reports includes the following recommendations to improve the program: (1) develop a science plan for the program; (2) examine the entire review process for the EMSP with the goal of increasing its transparency and technical credibility; (3) find a solution to the problem of not being able to "forward fund" projects at national laboratories, and fully fund all awards in the first year; (4) establish an EMSP program director responsible for management of the program who reports directly to the Under Secretary of Energy; (5) convene an independent review panel to review the performance and effectiveness of the program; and (6) convene annual workshops, seminars, and symposia to help facilitate information flow and stimulate new research ideas.

## PEER REVIEW IN ENVIRONMENTAL TECHNOLOGY DEVELOPMENT PROGRAMS: THE DEPARTMENT OF ENERGY'S OFFICE OF SCIENCE AND TECHNOLOGY (NRC, 1998)

This report provides an overview of an effective peer review program and its use in R&D decision making. In particular, the report focuses on how peer review can be used to evaluate the technical merit of environmental remediation technologies at various stages of development from basic research through demonstration to deployment. The report includes recommendations on how the Office of Science and Technology (OST) in DOE's Office of Environmental Management (EM) could improve its peer review process, and the linkage of peer reviews to its decision-making processes.

## DECISION MAKING IN THE U.S. DEPARTMENT OF ENERGY'S ENVIRONMENTAL MANAGEMENT OFFICE OF SCIENCE AND TECHNOLOGY (NRC, 1999a)

This report examines the prioritization and decision-making processes of DOE-EM's OST. The committee found that OST's decision process is closely linked with the DOE-EM organizational structure, institutional procedures, and program management. The committee framed its major recommendations around the four decision process issues raised in the study charter: appropriateness and effectiveness of OST's decision-making process, appropriate technical factors and the adequacy with which they can be measured, role and importance of effective reviews, and program challenges and measures of success. Specific recommendations include (1) OST should use the best available information on DOE-EM site technology needs as a guide for tailoring program goals; (2) the decision process should be structured using quantifiable attributes wherever applicable but also should allow for managerial flexibility; (3) OST should use the minimum number of stages and gates needed to track a project and should use peer reviews; and (4) the gate reviews of stage-and-gate tracking system should also assess estimations of cost, risk, and schedule.

## GROUNDWATER AND SOIL CLEANUP: IMPROVING MANAGEMENT OF PERSISTENT CONTAMINANTS (NRC, 1999b)

This report advises DOE on technologies and strategies for cleaning up three types of contaminants in groundwater and soil: (1) metals, (2) radionuclides, and (3) dense nonaqueous-phase liquids (DNAPLs), such as solvents used in manufacturing nuclear weapons components. Metals and DNAPLs are common not only in the weapons complex but also at contaminated sites nationwide owned by other federal agencies and private companies. They have proven especially challenging to clean up, not just for DOE but also for others responsible for contaminated sites. The report makes a number of recommendations, including the following: (1) in situ remediation should receive a higher priority in the Subsurface Contaminants Focus Area (SCFA); (2) SCFA should work more closely with technology end users in setting its overall program direction; (3) SCFA should sponsor more field demonstrations; and (4) DOE managers should reassess the priority of subsurface cleanup relative to other problems and, if the risk is sufficiently high, they should increase remediation technology development funding accordingly. Although the recommendations are designed for DOE, the bulk of the report will be useful to anyone involved in the cleanup of contaminated sites. The report also contains reviews of regulations applicable to contaminated sites, the state of the art in remediation technology development, and obstacles to technology development that apply well beyond sites in the DOE weapons complex.

## AN END STATE METHODOLOGY FOR IDENTIFYING TECHNOLOGY NEEDS FOR ENVIRONMENTAL MANAGEMENT, WITH AN EXAMPLE FROM THE HANFORD SITE TANKS (NRC, 1999c)

While DOE has a process based on stakeholder participation for screening and formulating technology needs, it lacks transparency (in terms of being apparent to all concerned decision makers and other interested parties) and a systematic basis (in terms of identifying end states for the contaminants and developing pathways to these states from the present conditions). The primary purpose of this study is to describe an approach for identifying technology development needs that is both systematic and transparent to enhance the cleanup and remediation of the tank contents and their sites. The committee believes that the recommended end-state-based approach can be applied to DOE waste management in general, not just to waste in tanks. The approach is illustrated with an example based on the tanks at the DOE Hanford Site in Washington state, the location of some 60 percent (by volume) of the tank waste residues.

## ALTERNATIVE HIGH-LEVEL WASTE TREATMENTS AT THE IDAHO NATIONAL ENGINEERING AND ENVIRONMENTAL LABORATORY (NRC, 1999d)

This report assesses the technical alternatives to calcining of high-level waste (HLW) at the Idaho National Engineering and Environmental Laboratory (INEEL). The calcination process injected waste into a fluidized bed at elevated temperatures to evaporate the water and decompose other material into calcine, a granular ceramic. The calcine was sent to storage in partially buried stainless steel bins enclosed by a concrete vault. As tanks were emptied of HLW, they were used to store liquid waste. The liquid is mixed transuranic (TRU) waste high in sodium, referred to as sodium-bearing waste (SBW). Some of the SBW has been calcined, and for several decades, R&D activities at INEEL have studied technical alternatives for the future remediation, storage, and ultimate disposition of HLW calcine and SBW. The committee concluded that the interim storage of calcine in the bins should be maintained until it becomes clear (1) where the material can be sent, (2) what disposal form(s) are acceptable, and (3) that an approved transportation pathway to a disposal site is available. The committee also concluded that DOE should solidify the SBW as soon as practicable and recommends that solidification options other than calcination be identified. The committee also concluded that a major consideration in deciding how (and whether) to process any radioactive waste for long-term conditioning is that of the risks being added and/or mitigated.

## LONG-TERM INSTITUTIONAL MANAGEMENT OF U.S. DEPARTMENT OF ENERGY LEGACY WASTE SITES (NRC, 2000a)

This study examines the capabilities and limitations of the scientific, technical, and human and institutional systems that compose the measures that DOE expects to put into place at potentially hazardous, residually contaminated sites. The committee found that, at a minimum, DOE should plan for site disposition and stewardship much more systematically than it has to date. At many sites, future risks from residual wastes cannot be predicted with any confidence, because numerous underlying factors that influence the character, extent, and severity of long-term risks are not well understood. Among these factors are the long-term behavior of wastes in the environment, the long-term performance of engineered systems designed to contain wastes, the reliability of institutional controls and other stewardship measures, and the distribution and resource needs of future human populations.

## RESEARCH NEEDS IN SUBSURFACE SCIENCE: U.S. DEPARTMENT OF ENERGY'S ENVIRONMENTAL MANAGEMENT SCIENCE PROGRAM (NRC, 2000c)

The report provides an overview of the subsurface contamination problems across the DOE complex and shows by examples from the six largest DOE sites (Hanford Site, Idaho Engineering and Environmental Laboratory, Nevada Test Site, Oak Ridge Reservation, Rocky Flats Environmental Technology Site, and Savannah River Site) how advances in scientific and engineering knowledge can improve the effectiveness of the cleanup effort. The committee analyzed the current EMSP portfolio of subsurface research projects to assess the extent to which the program is focused on DOE's contamination problems. The committee also reviewed related research programs in other DOE offices and other federal agencies to determine the extent to which they are focused on DOE's subsurface contamination problems. On the basis of these analyses, the report identifies the highly significant subsurface contamination knowledge gaps and research needs that the EMSP must address if the DOE cleanup program is to succeed. The committee recommends that the subsurface component of the EMSP have the following four research emphases: (1) location and characterization of subsurface contaminants and characterization of the subsurface, (2) conceptual modeling, (3) containment and stabilization, and (4) monitoring and validation.

## LONG-TERM RESEARCH NEEDS ON RADIOACTIVE HIGH-LEVEL WASTE AT DEPARTMENT OF ENERGY SITES: INTERIM REPORT (NRC, 2000d)

The committee was asked to provide this interim report to help the EMSP develop a request for proposals (RFP) aimed at HLW management for fiscal year 2001. The committee identified broad research areas that would benefit from a basic science plan and concluded that the RFP should solicit research projects in the following four fields, in order of importance: (1) long-term issues related to tank closure and characterization of surrounding areas; (2) high-efficiency, high throughput separation methods that would reduce HLW program costs over the next few decades; (3) robust, high-loading, immobilization methods and materials that could provide enhancements or alternatives to current immobilization strategies; and (4) innovative methods to achieve real-time and, when practical, in situ characterization data for HLW and process streams that would be useful for all phases of the waste management program. The committee also provided recommendations on several programmatic issues: (1) EMSP should promote "needs driven" or "mission-directed" basic science supporting research on fundamental processes and phenomena with potential high-impact results; (2) EMSP should promote underlying science and technology parallel to baseline or programmatic approaches to enable HLW management efforts to be flexible in dealing with any unanticipated difficulties; and (3) EMSP investigators should interact with problem holders at the sites to learn about the nature of the problems to be solved. The committee plans to produce a final report with more detailed findings and recommendations in the summer of 2001.

## LONG-TERM RESEARCH NEEDS FOR DEACTIVATION AND DECOMMISSIONING AT DEPARTMENT OF ENERGY SITES: INTERIM REPORT (NRC, 2000e)

The committee was asked to provide an interim report that addressed the technical content of a fiscal year 2001 EMSP call for research proposals and made recommendations on the areas of research where the EMSP could make significant contributions to solving deactivation and decommissioning (D&D) problems and adding to general scientific knowledge. The committee identified three areas where EMSP-funded research could make significant contributions: characterization, decontamination, and remote systems. Within these areas, it made five recommendations: (1) basic research toward identification and development of real-time minimally invasive and field-usable means to locate and quantify difficult contaminants significant to D&D; (2) basic research that could lead to the development of biotechnological sensors to detect contaminants of interest; (3) basic research toward fundamental understanding of the interactions of important contaminants with the primary materials of interest in D&D projects; (4) basic research on biotechnological means to remove or remediate contaminants of interest from surfaces within porous materials; and (5) basic research toward creating intelligent remote systems that can adapt to a variety of tasks and be readily assembled from standardized modules. The committee also provided DOE with the following general advice on EMSP strategic planning: (1) avoid focusing too narrowly on site-specific problems; (2) develop a more comprehensive, coordinated, and specific definition of complex-wide D&D needs; (3) allow DOE contractors and Site Technology Coordinating Groups to contribute more toward identifying true R&D opportunities; (4) help develop a scientific basis for setting standards for the end states of D&D; and (5) consider further interdisciplinary collaborations among relevant disciplines. The committee plans to produce a final report in the spring of 2001, which will provide more detail on the recommendations and advice in the interim report.

## ALTERNATIVES FOR HIGH-LEVEL WASTE SALT PROCESSING AT THE SAVANNAH RIVER SITE (NRC, 2000f)

The original process developed to accomplish the processing of high-level radioactive waste salt solutions stored at the Savannah River Site was in-tank precipitation (ITP), which encountered unexpected problems. A primary alternative selected by the Savannah River Site was a variation of ITP, known as small tank precipitation using sodium tetraphenylborate (TPB) and a backup option, crystalline silicotitanate (CST) ion exchange process. Other options, eliminated by the Savannah River Site, include caustic side solvent extraction and direct grout. This report reviews both the selection process of the two primary alternatives, and the processing options themselves. The committee found that there are potential barriers to implementation of all the alternative processing options and recommends that the Savannah River Site proceed with a carefully planned and managed R&D program for three of the four alternative processing options (small tank precipitation using TPB, CST ion exchange, and caustic side solvent extraction) until enough information is available to make a more defensible and transparent downselection decision.

## NATURAL ATTENUATION FOR GROUNDWATER REMEDIATION (NRC, 2000h)

The term "natural attenuation" refers to the use of unenhanced natural processes for site remediation. The biological, chemical, and physical processes, such as biodegradation, take place in the subsurface and may transform contaminants to less harmful forms or immobilize them to reduce risks. This report takes a look at public concerns about natural attenuation, the scientific bases for natural attenuation, and the criteria for evaluating the potential success or failure of natural attenuation. The principal findings of the report are that natural attenuation is an established remedy for only a few types of contaminants, that rigorous protocols are needed to ensure that natural attenuation potential is analyzed properly, and that natural attenuation should be accepted as a formal remedy for contamination only when the processes are documented to be working and are sustainable. Where communities are affected by contamination, community members must be provided with documentation of these processes and given an opportunity to participate in decision making.

## IMPROVING OPERATIONS AND LONG-TERM SAFETY OF THE WASTE ISOLATION PILOT PLANT: INTERIM REPORT (NRC, 2000i)

This committee was asked to advise DOE on the operation of the WIPP and to provide recommendations on two issues: (1) a research agenda to enhance confidence in the long-term performance of WIPP and (2) increasing the throughput, efficiency, and cost-benefit without compromising safety of the national transuranic (TRU) program for characterizing, certifying, packaging, and shipping waste to WIPP. This interim report provides DOE with recommendations on research to enhance confidence in long-term repository performance and improvements to the national TRU program. The committee recommended that DOE develop and implement a plan to sample oil-field brines, petroleum, and solids associated with current hydrocarbon production to assess the magnitude and variability of naturally occurring radioactive material in the vicinity of the WIPP site; eliminate self-imposed waste characterization requirements that lack a legal or safety basis; derive a more realistic gas generation model; consider cost-effective ways to improve the reliability and ease of use of the Transportation Tracking and Communication System; and develop tools for maintaining information needed to respond to a WIPP transportation accident.

## DISPOSITION OF HIGH-LEVEL WASTE AND SPENT NUCLEAR FUEL: THE CONTINUING SOCIETAL AND TECHNICAL CHALLENGES (NRC, 2001)

The concept of geological disposal is not new, yet many national programs have been faced with significant challenges siting a geological repository and emplacing spent nuclear fuel and HLW in it. This study, authored by a committee of experts from seven countries, addresses some of the challenges that national programs have confronted or are currently dealing with. The committee concluded that focused attention by world leaders is needed to address the substantial challenges posed by disposal of spent nuclear fuel and HLW. In addition, the biggest challenges in achieving safe and secure storage and permanent waste disposal are societal. Technically, there are only two feasible options: (1) storage on or near the Earth's surface and (2) placement in deep underground repositories. After four decades of study, the geological repository option remains the only scientifically credible, long-term solution for safely isolating waste without having the rely on active management. Furthermore, although there are still some significant technical challenges, the broad consensus within the scientific and technical communities is that enough is known for countries to move forward with geological disposal. This approach is sound as long as it involves a step-by-step, reversible decision-making process that takes advantage of technological advances and public participation.

# APPENDIX G

# LIST OF ACRONYMS AND ABBREVIATIONS

| | |
|---|---|
| AAAS | American Association for the Advancement of Science |
| BER | Office of Biological and Environmental Research (DOE) |
| BES | Office of Basic Energy Sciences (DOE) |
| COSEPUP | Committee on Science, Engineering, and Public Policy (National Academies) |
| CRESP | Consortium for Risk Evaluation with Stakeholder Participation |
| D&D | Deactivation and Decommissioning |
| DOD | Department of Defense |
| DOE | U.S. Department of Energy |
| EM | Office of Environmental Management (DOE) |
| EMAB | Environmental Management Advisory Board (DOE) |
| EMSP | Environmental Management Science Program (DOE) |
| EPA | U.S. Environmental Protection Agency |
| EQ | environmental quality |
| EMAB | Environmental Management Advisory Board |
| ERPS | Environmental Restoration Priority System |
| GAO | U.S. General Accounting Office |
| HLW | high-level waste |
| INEEL | Idaho National Engineering and Environmental Laboratory |
| IRI | Industrial Research Institute |
| LLW | low-level waste |
| MOU | memorandum of understanding |
| MD | Office of Fissile Materials Disposition (DOE) |
| MTHM | metric tons heavy metal |
| MLLW | mixed low-level waste |
| NE | Office of Nuclear Energy, Science and Technology (DOE) |
| NERAC | Nuclear Energy Research Advisory Committee |
| NERI | Nuclear Energy Research Initiative |

| | |
|---|---|
| NRC | National Research Council |
| NSF | National Science Foundation |
| NWTRB | Nuclear Waste Technical Review Board |
| OMB | Office of Management and Budget |
| ORNL | Oak Ridge National Laboratory |
| OST | Office of Science and Technology (DOE-EM) |
| OTA | Office of Technology Assessment |
| R&D | research and development |
| RCRA | Resources Conservation Recovery Act |
| RFP | request for proposals |
| RW | Office of Civilian Radioactive Waste Management (DOE) |
| SBW | sodium-bearing waste |
| SERDP | Strategic Environmental Research and Development Program (DOD) |
| SC | Office of Science (DOE) |
| SCFA | Subsurface Contaminants Focus Area |
| SLC | Strategic Laboratory Council |
| SNF | spent nuclear fuel |
| TD&T | Technology Development and Transfer |
| TRU | transuranic waste |
| TSCA | Toxic Substances Control Act |
| USNRC | U.S. Nuclear Regulatory Commission |
| WAG | Washington Advisory Group |
| WIPP | Waste Isolation Pilot Plant |
| WPRS | Work Package Ranking System |

To

Dansiy the angel
at the gym.

Michael

Sept 06/2018

# The Angel with a Broken Heart

**Leslie Michael**

A Great Little Publishing Company Book

North Vancouver

All the characters and events portrayed are taken from the Holy Bible. Any resemblance to persons living or dead is purely coincidental.

Published by
> The Great Little Publishing Company
> P.O. Box #87, 340 Seymour River Place
> North Vancouver, B.C., Canada. V7H 1S8
> May 2010

To obtain copies visit www.GreatLittlePublishing.com

Library of Congress Catalog Number/
Library and Archives Canada Cataloguing in Publication

Michael, Leslie, 1937-
> The angel with a broken heart / Leslie Michael. 1st Edition

ISBN 978-0-9737224-6-8

> I. Title.

PS8626.I15A76 2010     C813'.6     C2010-903560-7


First edition: May 2010   (Hardcover)

Printed in the United States of America
> 12  11  10  9

# The Angel With A Broken Heart

## Leslie Michael

# Acknowledgments

With love and gratitude to my daughter Anne for chal lenging me to write this book.

Heartfelt thanks to two good friends, Kristin Bregoliss and Maria, who never doubted my ability to write and for their implicit faith in me and their unfailing support and en couragement.

I am grateful to my editor, Viktoria Cseh, who advised me on the revisions and who, in her notes and conversations, always found a way to be exacting, thorough, and yet encour aging.

Numerous friends assisted me and insisted that I finish this book. I thank the librarians at the Douglas College Li brary, Coquitlam, British Columbia, Canada, who readily and happily directed me to books and other references.

To Gregg Patterson, whose knowledge of history pro duced interesting and intelligent discussions that gave me a great deal of insight as how to proceed with certain chapters of the book. I thank him for his kindness, encouragement, and friendship.

This book could never have been written without my recourse to the inspired word of God as laid down in the Holy

Bible. Neither would it have been possible without the work of all the authors and historians who have gone before me.

Though this book is built on the foundations of history, it is also the result of my imagination. I am solely responsible for any intentional or accidental straying away from history that the world has, without question or reservation, come to accept as fact.

It is for this precise reason that I set about to find out how, after the creation of the world, the creation of the first man and woman, a talking serpent (Satan) came to be in Par adise — the Garden of Eden. The story that follows is mine and mine alone. I accept full responsibility for it.

A number of books introduced me to the fall of Lucifer, the most magnificent angel, created from fire and embedded with precious stones, and the his seduction of Eve. Chiefly among them are the epic poem *Paradise Lost,* by John Milton and *Patriarchs and Prophets,* by E. G. White.

I acquired additional information about the life of Mary and Joseph before they were betrothed from various ar ticles, magazines, and books as nothing is mentioned in the Holy Bible. These are too numerous to list.

Pounding away at the keyboard, buried in one's own thoughts, and living in a very private world would have been unthinkable if it were not for family and friends who not only helped me find the strength and courage but who also filled my life with joy, laughter, and happiness in what would other wise, have been a very lonely time.

# Dedication

This book is dedicated, with great respect, to the people of the Middle East — the past, the present, and to come.

The Jews and the Arabs — two of God's very special people, descended from a common father, Abraham — were both promised greatness and blessings. Abraham hailed from the city of Ur in present day Iraq, while Hagar, the maidservant of Sarah and the surrogate wife of Abraham, came from Egypt. The Holy Bible gives vivid accounts of the open and unguarded border between these people. Abraham crossed into Egypt because of a famine that raged on his side. Joseph, who had been sold into slavery, witnessed his father and brothers cross over a few times. The gentle and kind daughter of the Pharaoh braved the harsh law of her ruthless father when she took the baby Moses as her own child. Her kindness and compassion for the child, saved for the Hebrews, the man who would eventually become the great lawgiver and who would be the one to lead the Israelites from slavery in Egypt to the Promised Land.

Centuries later King Solomon's marriage to another Pharaoh's daughter brought him great wealth. And some five thousand years later Mary, Joseph, and the child, Jesus, fled from Judea into the sanctuary of Egypt.

Yet, over the centuries, they have fought savage wars,

brother against brother, failing to heed the divine promises of God, that both nations would rise to greatness. And as the intensity of the intolerance and hatred continues between these two peoples, I remind them, most humbly, that God does not speak vainly nor does he make idle promises. I earnestly implore them, not to doubt as Abraham did, but, in the face of serious difficulties, misunderstandings, distrust, and hostility, to be steadfast in their faith and trust in God.

This book, then, is a prayer for them: that Heaven will touch earth and hope will be renewed; That Almighty God, in His infinite goodness and mercy, will grant them peace, will protect them always, will keep them safe from all disquiet, and will calm the storms that rage in their hearts and in their minds.

# Introduction

A long time ago and far away, a woman waited — and waited. She loved her husband dearly and was saddened by the fact that she was unable to bear him children. She was well past her child-bearing years and concluded that she was cursed with an empty womb. Nevertheless, she was still de termined that her husband would have descendants to per petuate his name. She made a momentous decision.

Sarah, the woman, told her husband, Abraham, that the Lord had kept her from having children. She brought her maidservant, Hagar, to Abraham and suggested that he should take on a surrogate wife so that he may have sons through her. Abraham complied from the earnest desire of the fulfillment of God's promise that his descendants would be as numerous as the stars in heaven and countless as the grains of sand on a seashore. Hagar was from Egypt and she had cho sen to travel with Abraham's camp when he passed through that country and to serve as Sarah's maid.

Hagar soon found herself with child and looked scorn fully on her mistress. Perhaps she upbraided Sarah for her barrenness and boasted of her prospect of bringing an heir to Abraham and thus fulfilling God's promise to him. Or maybe Sarah was filled with anger and jealousy and blamed Abra ham for Hagar's insolence and disdain. Abraham reminded her that Hagar was her maid and told her to do whatever she wishes. Hagar was so despised by Sarah that she fled from her.

While she wandered in the wilderness on her way back to her own country, Meteron, an angel of the Lord, the very first angel to be mentioned in the Sacred Scriptures, found her by a fountain of water. Addressing her by name and as Sarah's maidservant, Hagar told him she was fleeing her mis tress because she could not tolerate the harsh way in which she had been treated. The angel assured her that God had heard of her affliction and instructed her to go back to her mistress and to submit to her abusive treatment. Hagar was obedient to the good council and the angel assured her of the mercy that God had in store for her and for her offspring; that they also would be too numerous to count. He foretold that she would bear a son and his name will be Ishmael and that he will dwell in the presence of his brethren. Hagar pro claimed the name of the Lord and was gracious and grateful for the help God had given to a woman in distress. Hagar did as the angel had commanded her. She returned to Sarah, her mistress, and was submissive to her and eventually gave birth to her son, Ishmael.

God then spoke to Abraham assuring him that his wife, Sarah, would be blessed with a son and he shall be called Isaac. He promised him that His covenant will be with him and his generations; that he would be the father of many na tions; that He will make him exceedingly fruitful and that He will give unto him and his generations after him, all the land of Canaan from the Wadi of Egypt to the Great River, the Euphrates, and that He will be their God forever. Abraham doubted the word of God as he was 100 years old and his wife, Sarah, was 90.

Abraham still clung to the belief that it would be through Ishmael that God's divine purposes would be fulfilled. With great affection for his son, he interceded for him, exclaiming, "O that Ishmael might live before Thee!" God assured him that He will bless him as well and make him fertile and will multiply him exceedingly. He will be the father of twelve chieftains and will make of him a great nation.

Abraham then remembered the covenant he made with God that all male children were to be circumcised and so he took all male members of his household for this ritual. Thus Abraham and his son, Ishmael, were circumcised on the same day. Eventually, and according to God's faithful promise, Sarah bore Abraham a son and named him Isaac.

The birth of Isaac brought joy and gladness to Abraham, Sarah, and members of their household. Unfortunately for Hagar, this was an event that would see her fondly cherished hopes and dreams dashed. As the two sons grew, Sarah noticed that the son whom Hagar the Egyptian had borne to Abraham was playing with her son, Isaac. She demanded that Abraham drive that slave and her son away as she desired that no son of that slave would gain any inheritance with her son, Isaac. Abraham was greatly distressed, especially on account of his son, Ishmael. But God told him to heed the demands of his wife, Sarah, and assured him that He will make a great nation of the son of the slave woman since he was also his offspring.

Early the next morning, Abraham got some bread and a skin of water, gave them to Hagar and sent her away with her son, Ishmael. As she wandered aimlessly in the wilderness, the water was used up. So she put the child down under a shrub and prayed she will not watch the child die. As she sat opposite him, the child began to cry. God heard the boy's cry and Meteron, an angel of the Lord, called to Hagar from heaven telling her not to be afraid for God had heard the boy's cry. He then commanded Hagar to arise and lift the boy and hold him by the hand for God planned to make of him a great nation. Her eyes were then opened and she saw a well of water. She went and filled the skin with water, let the boy drink and continued on her way.

God was with the boy as he grew up. He lived in the wilderness and became an expert bowman. With his home in the Paran, his mother got a wife for him from the land of Egypt.

God then put Abraham to the test because he had doubted the Word of the Lord. He was chosen to be the father of the faithful and to stand as a shining example for future generations. But he was sadly lacking in faith. He had demon strated this many years ago in Egypt by concealing the fact that Sarah was his wife and persuaded her to masquerade as his sister. He was unfaithful in his marriage by having a child through Hagar, his wife's maidservant. God now commanded Abraham to take his son, Isaac, up a mountain and sacrifice him as a burnt offering. On reaching the place that God had told him, Abraham built an altar, arranged the wood on it, bound his son, and put him on top on the altar. He then reached for his knife to slaughter his son, but angel Meteron called to him and told him not to lay a hand on the boy.

The great act of faith and obedience of Abraham stands like a beacon, illuminating the path of God's faithful. He had chosen not to offer any excuses or reason that the slaying of his son would contradict God's promise to him. He was im plicitly obedient to God's command.

It must be noted that the two brothers, Ishmael and Isaac, often played together as children and that they were both saved by the angel of the Lord. As they went their sepa rate ways, their races increased and multiplied with no one paying any attention to the word of God. Peace, tranquility, and harmony should have prevailed in their Holy Lands. In stead, it is anger, suspicion, hatred, and war that dominate their regions.

The guns are never silent and the skies rain death. A great tragedy is unfolding. Since the dawn of time, men have sought peace, but power-hungry men and military alliances have served only to disturb the peace, paving the path to war, and leaving in its wake untold sorrow and suffering. If the people of the Middle East do not come up with a just and more equitable solution Armageddon, the final battle, will be upon us. The intelligence and the hard work of these people, their spiritual strength and faith, ought to be channeled into an earnest desire for peace that would raise them up from the ever-fearful conditions that they live in and turn it into a posi tion of dignity.

In spite of centuries of violence, hatred, intolerance, and armed conflicts, both nations are still standing. It is only through a meeting of minds that a common ground would be found to bring peace and harmony to their troubled lands. It is only then that they would experience the joy that flows from hope.

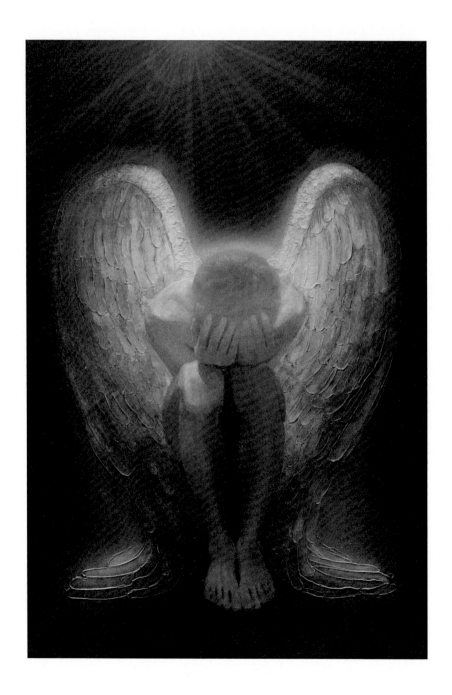

Adapted from original art by Liliya Popova

# The Angel With A Broken Heart

**Leslie Michael**

# Table of Contents

# List of Illustrations

*And it came to pass, when the Lord Jesus was born at Bethlehem of Judaea, in the time of King Herod, behold, Magi came from the East to Jerusalem, as Zeraduscht had predicted; and there were with them Gifts: gold, frankincense, and myrrh. And they adored him, and presented him their gifts.*

*Then the Lady Mary took one of the swaddling-bands, and, on account of the smallness of her means, gave it to them; and they received it from her with the greatest marks of honor.*

*And in the same hour there appeared to them an angel in the form of a star which had before guided them on their journey; and they went away, following the guidance of its light, until they arrived in their own country.*

*— from The Arabic Gospel of the Infancy of Jesus.*

# The Angel With

# A Broken Heart

Leslie Michael

*God revealing His divine plan to the angelic hosts*

# 1
# The Heavenly Council

In the beginning God saw the beauty, harmony, and the glorious peace that existed in Heaven. In that uni versal kingdom of His, He had placed magnificent super natural beings, called angels, who were radiant with light and could travel with unimaginable speed to do God's bidding. They could also assume any shape, form, or size and moved freely among the stars, their movements being like heavenly bodies. They formed the government of God according to His law of love without being forced into obedience but rendering service freely and with a sense of love, not only for God but also for one another,

thereby, creating that tranquility, peace, and perfect har
mony.

But God also saw that the void below was a form
less wasteland. It was empty and darkness covered the
abyss. There was nothing in it. It was a formless mass of
utter chaos, emptiness, and confusion. It was barren, des
olate, foreboding, and a mighty wind swept over it. God
is a good and wise creator, a higher being of incredible
spiritual and intellectual qualities. He is omnipotent, om
nipresent, and omniscient. He is infinite and in His infi
nite wisdom, absolute power, and absolute goodness God
decided to bring order out of chaos. He would create
Earth – another world – and give Heaven to Earth.

God assembled together the multitude of His an
gelic hosts and explained His divine plan to the heavenly
council. He also let it be known that in addition to creat
ing a new world He would also create man. He would be
lower in status than the angels but created in the image
and likeness of God. He also made it abundantly clear to
the heavenly hosts that even though man would be creat
ed inferior to them they would be commanded to treat
him as an equal and honor him. Man would not be creat
ed to be trampled upon by superior forces, to be crushed

under the conqueror's heel, to be enslaved and robbed of his possessions. He reminded the angelic hosts that they were not created to rule the lesser beings.

Lucifer did not think it was a great idea or even a divine plan. He was the "Bearer of Light," the "Shining Star of Dawn," the "Son of the Morning." He was a Super natural Being, of dazzling brightness, of majestic, awe some beauty with the gift of supreme power and wisdom. He was perfect in every way from the day he was created from fire and embedded with precious stones. He was the most honored of God's angels, highest in power and glory among the angelic hosts.

It seemed to Lucifer that God was planning to cre ate a new world order and place in it a creature far below him and that he and the heavenly hosts would be com manded to pay homage to this inferior being. He had al ways lovingly acknowledged the supremacy and majesty of God. Now it seemed that instead of God creating order out of the chaos below, he was creating confusion and disharmony in Heaven above. Lucifer was the most loyal lover of God and the most amazing and striking image of Lucifer was his great love and deep devotion to his Cre ator. He often prostrated himself before God and poured out his love and adoration. Lucifer was also loved and re spected by his fellow angels. But God was now demand ing the unthinkable: creating a creature called "man" and ordering the angelic hosts to bow in reverence to that man whom God said would be the noblest of all His creations.

Lucifer vowed he would not do this. He could not possibly imagine that anyone could be more glorious than he was. He was too proud and that cardinal sin prevented him from perceiving the Divine image of God which would be placed in man. That sin of pride prevented Lu cifer from understanding God's plan of creating and giv ing man, just like He had done with Lucifer and the an gelic hosts, the ultimate gift of establishing a loving rela tionship with his Creator; of sharing Heaven with Earth. God was also giving the angels the opportunity to be kind and loving to someone who would be their inferior. Lu cifer failed miserably to understand that his love for God would be reflected in his love for someone not his equal. The sad lack of humility led Lucifer to self-indulgence and utter dissatisfaction began settling in. He was not content with being the most magnificent of all angels but also coveted the homage due only to God. That inflated sense of entitlement placed him in the dreadful position of choosing between the Divine Order, the Angelic Order, or the Human Order. He made a fateful choice.

Calling all the angels together Lucifer held a full council. Like a coach before an important game he ex plained the situation. With head held high and eyes blaz ing he made it clear that to be weak and indecisive was miserable and to do nothing was not an option. After all the faithful services they had rendered it seemed that they had been mere slaves. God had turned His back on the angels and he called for vengeance. He assured them that he should not and would not bow to man. He felt far su perior and was convinced that it should be man who should pay homage to him. This great feeling of self-im portance convinced him this was a terrible injustice not only to him but to the other angels as well. They should feel free to claim their rights and not be deluded slaves bowing to an inferior being. Other proud angels who shared Lucifer's thoughts, ambitions, and visions cheered him on.

Archangels Michael, Gabriel, Raphael, and Me teron were very perturbed over this unhappy situation. Michael, whose very name means "Who is like God," was the Prince of the angelic hosts, the guardian of the keys of

Heaven and God's vice regent. He was patient and merci
ful. He was chivalrous and the defender of the universe
and Supreme Commander of the angelic legions. God
committed to him twelve legions, each consisting of
144,000 angels. Each had a divisional commander and it
was their duty to maintain peace, law, and order in the
universe, reporting directly to Michael.

Gabriel was the Chief Justice of the universe. He
was highly intelligent and endowed with great wisdom
and knowledge of the heavens. He was holy, just, and
truthful. He had under him angels specially trained in in
telligence gathering. They traveled the universe gathering
information and communicating with other angels in far-
flung corners of the heavens. All the information they col
lected would then be transmitted to Gabriel and, should
the need arise, Gabriel himself would journey to any cor
ner of the universe to explain the laws of God and dis
courage any rebellion. His shuttle diplomacy proved in
valuable to the Creator.

Raphael was the Chief Medical Officer. Gifted with
the knowledge of science he was a traveling companion to
the sick and infirm, healing many on their way and pro
tecting others. He was the angel of prayer, love, joy, and

light. He was one of the chosen angels who presented and recorded prayers before God. He was particularly kind to sinners, especially those who returned to righteousness; there was always great rejoicing in Heaven when a sinner repented.

Meteron was the merciful one. A kind-hearted, loving, and compassionate angel, he was always there to comfort the afflicted.

The four archangels were the guardians of the throne of God and represented perfection or totality as in the four cardinal directions. These four rulers or princes of the angelic hosts met with Lucifer and his rebellious followers, Belial, Asmadeus, and Mammon. They reminded them that God was the Supreme Being, the creator of the universe, the almighty and all powerful and that it was His prerogative to do whatever He thought was best. They reminded the dissatisfied angels of God's wisdom and justice and did their best to reconcile them to God's divine plan. God, they reiterated, was loving and merciful and would forgive them for their rebellious nature. They placed before them the inevitable and terrifying consequences that would follow from their rebellion if the harmony in heaven was destroyed. They further advised that the unhappy angels should perhaps wait and see the kind of world God had in mind and what sort of creature they were to pay homage to. Lucifer was not impressed nor was he convinced. In his keen and analytical mind he was certain that a wait-and-see approach would be folly. In the distant mirror he saw that inaction and indecision would be fatal and a major disaster because that would give the opposing side precious time to rearm and regroup. Never theless, Lucifer promised that he would think it over and in the meanwhile a non-aggression pact would prevail.

Lucifer was greatly encouraged by the fact that one third of the angelic hosts had sided with him. He was also aware of the rumors that many legions of angels in far-flung corners of Heaven were not entirely satisfied with the reign of God and would seize this revolt as an oppor tunity to join forces with him in a rebellion against God. And that gave him a sense of strength and invincibility. After much discussion with his commanders over the im pending battle and military strategy, Lucifer amassed the rebellious angels on the western side of Heaven and launched a surprise attack. He was fully aware that the eastern side had far more open spaces and was virtually indefensible. He felt positive that this strategic move would prove him victorious and eventually he would push Michael and his loyal angels over the edge and into the abyss below.

Michael was an angel of honor and always kept his word. He never ever dreamed that Lucifer would be a liar, would resort to deceit and would dishonor the truce—the waiting period that they had both agreed upon. It would be impossible to describe the hurt, the heartache, and the sorrow that Michael felt on learning that Lucifer had de clared his independence from God—a God of infinite love, warmth, and tenderness. Lucifer was now Satan—an adversary of God, and he had fired the first shot in the battle for Heaven.

Lucifer, the magnificent

**2**

# The Battle for Heaven

**S**atan ordered twenty-four commanders, 2,500 se

nior angels, and 280,000 other angels to cross over to the

eastern front in one massive assault against Michael and

his loyal angels. As Satan's troops lunged across the bor

der, Michael and his troops rallied as quickly and ably as

they could amid the confusion and complete disorganiza

tion. The rumbling of fiery chariots rained fire unto fron

tier defense positions. Angels flying at incredible speed

conducted both strategic and tactical raids against numer

ous targets with frightening success. The loyal angels

could hardly believe this battle was taking place. The

sheer weight of numbers, the shock of betrayal, the devas tating element of surprise, and the heavy casualties took their toll. The loyal angels began to withdraw.

Satan was flushed with victory. He decided that Scowmo, the headquarters of Michael, was not yet the pri mary target. He would tantalize, tease, and humiliate him by first surrounding Gralindrag, a place that commanded a strategic location. He ordered General Moloch to ad vance, surround the area but not take it. Moloch was an angel whose loyalty to Satan had never flagged and who never questioned an order. He was ordered instead to ha rass, demoralize, and bring its population to its knees through constant bombardment from the air. Satan fig ured that a victory here without any loss would boost the morale of his bad angels.

Unfortunately, this strategy worked against Satan. The besieged inhabitants of Gralindrag persevered and their survival became the cornerstone of all future resis tance. Under the command of Archangel Gabriel the loyal angels were able to launch a counteroffensive and recap ture vital areas, pushing General Moloch back. Gabriel was then able to bring fresh supplies to the battered area.

Satan's fatal decision to divert his troops to Galin

drag enabled Michael to fortify the defenses of Scowmo. The inhabitants turned it into a fortress surrounded by trenches and anti-chariot ditches. Countless obstacles were placed in the surrounding areas and anti-air defens es ringed the whole place. In addition, Archangel Michael recalled forces from far-flung posts in the East to bolster the number of troops.

Satan was horrified. He was always of the opinion that these far-flung outposts in the East were not satisfied with God's reign and he had hoped that they would join him, thus compelling Michael to fight a war on two fronts.

With more than seventy divisions, 1,700 fiery chari ots, and more than 1,000 combat pilots, General Mammon launched the attack on Scowmo. Satan knew that Mam mon was another angel whose loyalty and abilities were unquestionable and who would restore order and disci pline once Scowmo was captured. He was also sure that he would suppress any rebellion and set up the savage defenses that would be needed. The fighting was fierce, but the loyal angels under Archangel Raphael held their ground and the stubborn resistance inflicted heavy losses on General Mammon's forces. The situation looked very grim but with the passage of time, reinforcements were sent to Raphael.

General Mammon was now on the defensive. He and Satan realized that Scowmo definitely had to be taken for military and psychological reasons. The second offensive was now being planned.

Reinforced by more than 200 fiery chariots and 90,000 angels, Archangel Raphael was now in a better position to face the mighty force of General Mammon, stop a large chariot division, and prevent an enemy breakthrough. It was now obvious to Satan and his other high-ranking officers that the plan to take Scowmo was going to end in failure. Satan's dream of winning the battle for Heaven with a rapid, quick-fire, lightning strike was going to end in disaster. Buoyed by the fact that Satan and his minions were on the defensive, the Supreme Commander, Archangel Michael planned a counter-offensive. The plan was not only to drive Satan's forces from Scowmo but to attack the enemy along the entire front line.

Commanding the angelic forces in the North was Archangel Gabriel who outflanked and encircled General Asmadeus, inflicting heavy losses. Archangel Raphael, in the middle, stunned General Azazel and caused him to rapidly retreat leaving behind valuable mil

Archangel Uriel in the South pushed Gene

his forces all the way back to the starting pla

Satan began to blame his commanding officers for their failure to win a decisive and quick victory. He re placed many of them with less experienced officers. By saving Scowmo and halting any further advances by Sa tan's forces the loyal angels gained a great victory, inflict ed heavy losses, and destroyed or captured much enemy arsenal. Scowmo had proved to be harder to conquer than anticipated. In the meanwhile, Galindrag had begun re ceiving supplies and no matter how inadequate and spo radic they were the loyal angels remained defiant, fight ing valiantly to hold the area against Satan and his gener als who were absolutely stunned by the courage, the sheer determination, the fighting ability, and the loyalty of Michael's angels.

Michael, pouring over battle plans with his com manders, quickly realized that there was a major over sight on his part. In his zeal to push the invaders back, he had overlooked an area, though of no significant military importance, would still be a great psychological advan tage to Satan. That area was Michelograd and it was named in his honor. He instinctively knew that Satan would assemble his mighty forces to capture it and a vic tory there would boost the morale of Satan and his fol lowers. Conversely, a loss there would demoralize his own forces. Therefore, Michelograd had to be defended at all costs. Michael also realized that evacuating that place would prove more harmful than good. He was confident that his forces would fight tooth and nail to save an in habited place rather than an abandoned area. To defend this city he chose the young, courageous, and brilliant commander, Archangel Meteron.

Meteron encouraged everyone to apply themselves to the defense of that city, with every street to be barricad ed and every block to be turned into an impregnable fortress. Facing Archangel Meteron was General Sulaup, whom Satan had ordered to capture Michelograd at all costs. With more than 1,200 airships he pounded the city, inflicting heavy casualties. But Meteron and his angels hung in there doggedly in their defenses. More than 10,000 artillery weapons opened fire followed by 650 fiery chariots and a million combat angels. Every square mile was fought for savagely. The relentless struggle between the two forces amid the desolation and destruction knew no bounds. What was gained by Sulaup in one day was eventually regained by Meteron the next. It now became clear to Satan that Michelograd was not going to fall easi ly.

Sulaup's forces were totally exhausted and bat tle-weary. Meteron, with more than a million angels, 13,000 artillery guns, 875 fiery chariots, and 10,000 air ships, launched his counteroffensive. With **M-I-C-H-A-E-L** as their battle cry, they surged forward. His airships pounded Sulaup's defensive positions, the big guns rain ing destruction on the ground forces while fiery chariots wreaked havoc with lightning and thunderbolts. On the southern flank Meteron's forces had embarked on a wide sweep and had trapped the enemy. They had executed a major offensive by not only trapping the enemy but also encircling them. Archangel Meteron knew he was on the verge of a major victory; not by just defeating the enemy but with the ability to throw the invaders out of Heaven. Being a kind and compassionate commander he sent two of his officers across the battle-scarred area and offered Sulaup the option to surrender. The terms were simple: They would stop fighting and surrender and in return he would guarantee that, as prisoners, no harm would come to them. He also asked that they repent and ask forgiveness for waging a war against God and instead trust in the goodness, the kindness, and the love of a merciful Creator.

General Sulaup did not give an answer as he was fully aware that Satan had ordered him to keep fighting and not take a single step backward. He merely stated that he was obliged to consult with Satan.

Satan was furious. He ranted and raved, swearing that no senior commanding officer of his was ever to surrender. The battle for Heaven would go on. He deceived himself into thinking that there was still hope. Other units in the surrounding areas would come to the aid of Sulaup and Meteron's offensive would eventually be crushed. The offer to surrender was rejected.

Meteron's guns opened fire. Fiery chariots, lightning, and thunderbolts rained destruction and havoc once more. General Sulaup now saw the futility of continuing the battle and surrendered.

Satan, whose mighty forces in the initial stages had launched an incredible lightning strike from the air and had followed that with his superior ground forces of fiery chariots, was now staring at defeat, destruction, and hu miliation. If the retreat of his forces continued it would prove to Archangel Michael that Satan's once superior forces had been fatally wounded at Michelograd.

To boost the sagging morale of his troops, Satan and the senior commander of his armed forces, Beelze bub, decided that there just had to be another offensive. Beelzebub was a devoted and dedicated follower of Satan and implicitly obedient. He firmly believed in the great future and destiny of Satan. He had no doubt whatsoever that he and Satan would teach Michael a lesson and show him that they were still a mighty force to be reckoned with. He thought and acted like a faithful believer who had just undertaken a pilgrimage and whose faith had been miraculously restored. He believed with all his heart that the recent defeats were only temporary setbacks and they would reverse the tide. He was fully convinced that they were still strong and united and would surely emerge invincible.

Satan rallying his rebellious forces to battle

Satan ordered them to advance toward a place called Skurk. It would be here that they would launch the offensive. The importance and the urgency of defeating the enemy at Skurk was repeatedly emphasized. These were exceptional and extraordinary times and he would need the unfailing, unquestionable devotion and chivalry of his men. He explained to them his daring plans. They would defeat Michael at Skurk and that victory would pave the way for a push eastward. What Satan conve niently failed to tell them was that intelligence reports had indicated that Michael had allies in the far-flung western areas. They were still loyal to God and they were preparing to open a second front which would necessitate Satan's fighting on two fronts. Before that could happen it was imperative that he defeat Michael at Skurk. Forces based in the West were called to the eastern front. He would throw everything he had at Michael. He was firmly convinced that they were still strong and united. So sure was he that he decided to spearhead the attack himself.

Archangel Michael viewed the situation calmly. His intelligence-gathering angels conveyed to him that Satan was going to lead the attack. It was time to meet him head-on in battle. He gave his assistants the honor of dressing him in his battle gear, complete with his armor plate. Resplendent in his battle uniform, his officers took him to the front line. His very presence there produced thunderous applause and cheers. His name, **MICHAEL! MICHAEL! MICHAEL!** thundered throughout the ranks of his angelic hosts.

He immediately ordered vast stretches of anti-chariot trenches to be dug. He rightly surmised where Satan would use his chariots and had anti-chariot guns placed in strategic locations. Anti-airship guns were also set in position. Archangels Meteron, Gabriel, Raphael, and Uriel were given command of various units along the front line to ensure that no enemy could break through, then encircle and attack them from the rear.

From intelligence reports Michael learned the exact time that Satan would launch the offensive. Before that could be put into effect Michael ordered a massive artillery bombardment. The result caused a tremendous adverse effect on the morale of Satan's forces for they realized immediately that their battle plans had been compromised.

After his initial shock Satan reorganized his troops and launched his great offensive. With air cover and an intense artillery barrage, more than 500 fiery chariots, backed by lighter ones and infantry behind the satanic forces, tried to break through on six different occasions. But each time they were repelled. Several times they tried to break through the flanks of the defenders but Michael's brilliant and courageous commanders stopped them in their tracks. Every ferocious attack was met with ferocious defenders inflicting heavy losses on enemy troops and equipment until it became clear that the strength and morale of Satan's forces had finally been broken.

That's when Michael launched his great counter-of fensive. He now had the initiative while his enemy's forces were in disarray. Flanked by his commanders and with more than 2,000 airships, 20,000 artillery guns firing thunderbolts, and a million troops on fiery chariots, Michael himself lead the assault. Satan's forces were even tually crushed.

As Satan and his followers stood at the very edge of Heaven, he finally came face to face with Michael. Their eyes met and Michael felt a tinge of sadness for him. Here was Lucifer—the most brilliant and magnificent of God's angels, whose very name meant, "The Bearer of Light," but whose pride hindered him from seeing God's divine plan. He had rebelled against a good, kind, and merciful God. Here he was now, a sorry figure of his former brilliant self—defeated and humiliated. Michael got a grip on himself. This was not the time for sentiments or niceties. With one last battle cry, **"C-H-A-R-G-E!"** Michael and his angelic hosts pushed Satan and his minions over the edge into the dark and empty void.

Leslie Michael

Lucifer cast into Hell

# 3

# In the Sea of Flames

**S**atan tumbled down and down and down into the deep, dark, and frightening abyss. It seemed to take forever. He was shocked, frightened, and bewildered. He kept tumbling down and down the bottomless pit until suddenly there appeared to be some light. He could not make any sense of it until he crashed into the sea of flames.

Recovering slowly, he looked around and realized he was not alone. A third of the angelic hosts who had re belled against God were also down with him. Although they were in a sea of flames they were not being burned. It occurred to him that they were spirits and, hence, could

not suffer physically. Even though they had been decisive
ly beaten in battle, nobody had died. They were now in
this horrid place of wrath and fury. Satan realized that his
sin of pride had indeed been the ultimate source from
which the others had risen. In his desire to be more im
portant, more attractive, and more magnificent than the
others he had failed to acknowledge God's love, His wis
dom, His mercy, and His supremacy. Satan had an exces
sive belief in his own abilities—even upon embarking
upon a futile and doomed attempt to compete with God
for honor and worship—and that had interfered and pre
vented him from recognizing the Grace and Goodness of
God. He was now doomed to be away from the presence
of a loving and merciful God. He was in a place totally
alienated from his Creator.

By the light of the flickering flames Satan observed
that the rebellious angels were glaring at him with open
contempt. They were aware that he had chosen to oppose
an omnipotent God and had ended up totally defeated,
humiliated, and impotent. Like any good commander Sa
tan knew that he should never show any sign of cow
ardice or concern in the presence of his troops. Putting on
a brave face he assured them that vengeance would be his

and they would soon regain Heaven. God, he told them, was just another angel who could, with the proper planning, be defeated. Satan, once so magnificent, was now completely in ruins. In a torrent of wrath and consumed with anger he sought revenge. His pride blinded him and instead of seeking forgiveness he was obsessed with vengeance.

The creation of the world

# 4
# The Seven Days

**G**od in Heaven felt great sorrow at the loss of his angelic hosts. He knew there was work to be done. He would bring order out of chaos by putting into action His original and divine plan of creating Earth and placing in it a man in His own imagine and likeness. His descendants would fill the void created by the rebellious angels.

On the first day God said, "Let there be light." And there was light. And God saw how good the light was. God then separated the light from the darkness. God called the light "day" and the darkness He called "night." Thus, evening came and morning followed—the first day.

In His very first command and in His infinite mer

cy God bestowed upon Earth the great wonder and bless
ing of light. Just like a new-born baby, the first thing that
the child notices is the light that surrounds his new world
—so different from the darkness of the womb. The beauty
and wonder of light will enable that child to observe
God's works and find glory in them.

Then God said, "Let there be a dome in the middle
of the waters to separate one body of water from the oth
er." And so it happened. God made the dome and it sepa
rated the water above the dome from the water below it.
God called the dome "the sky." Evening came and morn
ing followed—the second day. The firmament was created
to divide the waters in the oceans and the waters in the
clouds that come down as rain. The Lord, by His wisdom,
founded the Earth and established the heavens. By His
knowledge and understanding the depths broke open and
the clouds dropped their dew. The firmament, also called
"the heavens," was God's abode. The infinite distance be
tween the heavens and Earth spoke eloquently of the
supremacy, the glory, and the majesty of God.

Then God said, "Let the water under the sky be
gathered into a single basin so that the dry land may ap
pear." And so it happened: The water under the sky was

gathered into its basin and the dry land appeared. God called the dry land, "the Earth," and the basin of water He called, "the sea." God saw how good it was. And God said, "Let the Earth bring forth vegetation, every kind of plant that bears seed and every kind of plant that bears fruit with its seed in it." And so it happened. God saw how good it was. Evening came and morning followed—the third day.

God turned His attention now to man's survival. He had already created light. The firmament had been placed, separating the waters from the sea and the rain from the sky. The sea and land were two separate entities, making sure that the waters from the sea did not over whelm the dry land. The Earth had immediately become fruitful and it brought forth grass and herbs. Fruits and vegetables were perpetuating themselves by the seeds within them. God, in His infinite goodness, was providing for man and beast even before they were created.

Then God said, "Let there be lights in the dome of the sky to separate day from night. Let them mark the fixed times, the days and the years, and serve as luminar ies in the dome of the sky to shed light upon the Earth." And so it happened. God made the two great lights, the greater one to govern the day and the lesser one to govern the night; and He made the stars.

God set them in the dome of the sky to shed light upon the Earth, to govern the day and the night, and to separate the light from the darkness. God saw how good it was. Evening came and morning followed—the fourth day.

Then God said, "Let the waters teem with an abundance of living creatures and on Earth let birds fly beneath the dome of the sky." And so it happened. God saw how good it was and blessed them, saying, "Be fruitful and multiply and fill the waters in the seas; and let the birds multiply on the Earth." Evening came and morning followed—the fifth day.

After five days of creation God turned His attention to living creatures and said, "Let the Earth bring forth all kinds of living creatures: cattle, creeping things, and wild animals of all kinds." And so it happened. God saw how good it was and blessed them and in His infinite wisdom commanded all living creatures to be fruitful and multiply so that His works would be preserved for all time.

Then God concluded His work by doing what He

had planned. Not just creating order out of chaos but placing on Earth a being that would have dominion over the fish of the sea, birds of the air, the cattle, all the wild animals, and all the creatures that crawl on the ground. So God created man in his own image; in the divine image of God he created him. God formed man from the dust of the ground and breathed into his nostrils the breath of life and man became a living soul. God said unto him, "See, I give you every seed-bearing plant all over the Earth and every tree that has seed-bearing fruit on it to be your food; and all the animals of the land, all the birds of the air, all the living creatures that crawl on the ground, and I give you all the green plants for food."

God, in His infinite wisdom, created man last. Therefore, there can be no doubt whatsoever in suspecting man of helping God in the creation of the Earth or even of him creating the world by himself. God then took stock of all his work and, behold, it was very good. And the evening and the morning were the sixth day.

On the seventh day He rested.

The creation of Eve

**5**

# The Garden of Delight

**G**od then planted a garden eastward in Eden and put man there—the man whom He had formed from the dust of the Earth. A river flowed through Eden to wa ter the garden and from Eden it parted into four tribu taries. The name of the first was Pison which encom passed the whole land of Havilah where there was gold. The gold of that land was excellent. Bdellium and lapis were also to be found there. The name of the second river was Gihon which encompassed the whole land of Ethiopia. The name of the third river was Tigris which flowed east of Assyria. The fourth river was the Eu phrates.

God called the man whom He created Adam and showed him his new home, a home adorned by nature. Heaven was his roof and the Earth was his floor. The shade from the trees was his resting place. It was a garden of delight and pleasure. Truly it was Paradise.

But Adam was alone.

Even though there were good and loyal angels in Heaven and rebellious ones below, man had nobody to talk to or share the delights of this wonderful place with. God, looking down kindly on his solitude, knew that it was not good for man to be alone and decided to make a suitable partner for him. He cast Adam into a deep sleep and while he was asleep he took a rib out of the man and formed a woman out of that rib. In His logic and perfec tion God created a woman. She was made out of the side of man—not to rule over him or be trampled under his feet—but to be at his side and be equal to him. Like a lov ing father He gave the woman to Adam who said, "This one is bone of my bone and flesh of my flesh and she shall be called woman. That is why a man leaves his father and mother and clings to his wife and the two of them become one body."

It was a marriage made in Heaven. Like the other

living creatures that were commanded by God to be fruit ful and to multiply the conjugal union was willed by God who saw that the Earth, His beautiful creation, was empty and should now be inhabited by human beings made to his image and likeness and perfect in every way.

While God had been busy creating the world, Satan had been languishing in Hell, still reeling from his devas tating defeat and his descent into a place of fear and ter ror. He had doubted the absolute power of God and had the temerity to challenge that omnipotence. He had failed miserably and now he found himself horrified and impo tent. Taking stock of his pathetic situation, it slowly dawned on him that this awful place was not really a prison. There were no walls, no fences, and no bars. He could move freely. He recalled how God had assembled the angelic hosts and told them of His plan to create order out of chaos and to place in it an inferior being that would be created out of dust. His curiosity got the better of him and he embarked on a determined and dangerous quest to find that place and that man.

He wandered though the universe and was abso lutely amazed at what he saw. The sun, the moon, the stars, and the planets were all in perfect harmony func tioning according to the divine and perfect law of God. Wow! When he beheld the heavens it was clear to Satan that this was truly the work of God's hands. He had set the sun, the moon, and the stars in place. He wondered why God should be so mindful of man or even care for him. He was lost amidst the splendor and beauty of the celestial bodies. Satan was about to give up his quest when he noticed one planet that was blue and green with varying hues. It was so beautiful that he just had to investigate further. Descending cautiously he saw an extraordinary garden that had a river with four tributaries flowing through it, watering the garden. He was impressed with the lush vegetation and trees bearing different fruits. He realized that such a magnificent place could never be left unguarded. The crafty, the cunning, the scheming, the rebellious angel devised a plan. His keen, analytical mind and superior intelligence told him that gardens have snakes. He also realized that he had superior physical powers and could appear in any form, shape, or size. Thus, with his swiftness and unimpeded intellectual energy he transformed himself into a serpent. With great trepidation, anxiety, and fear, Satan entered the Garden of Eden.

Leslie Michael

Satan looking for Adam

# 6
# The Serpent in the Garden

**S**atan crawled around the Garden of Eden. He be came more confident and daring as time went by. He was amazed and fascinated by the beauty, the peace, and tran quility around him. It was beautified and adorned and pleasing to the sight. It was enriched and resplendent with trees that yielded fruit which were tasty and good. Coiling himself around the trunk of a tree he saw Adam and wondered what in Heaven's name had God planned for this miserable creature whom He had created from the dust of the Earth?

And then he saw Eve. Her beauty and angelic form

took his breath away. Adam was created from dust but Eve was created from Adam. If Adam was created in the image and likeness of God from dust then Eve, created from Adam, was Heavenly perfection—so beautiful and perfect that he was sure even night's eyes took in her loveliness.

He was seething with anger. He couldn't believe that Adam, created from dust, could be given a pic turesque garden with everything for the good of man in it. God also loved him so much that he gave him domin ion over the birds in the air, the fish in the waters, and the beasts in the fields. The heavens and the Earth were also his and God had commanded the heavens to give him rain and the Earth to give him fruit. The sun and the moon were also created for him—the sun to give him light by day and the moon to shine by night. He was not lack ing in any comfort and still he had been given a beautiful woman to love and to cherish.

Satan was tortured by the grace and angelic beauty of Eve, knowing fully well that she was not his to enjoy and this torment aroused in him further hatred and envy. He was the most beautiful and brilliant angel and even though he was the greatest power ever to be overthrown, he was confident that he would stand up like a giant in the face of this terrible injustice. He still had a strong will to be revengeful and also the courage to never yield or submit to this outrage. He was seething with vengeance. He was fierce in his pride even though his ego was taking a terrible bashing. He decided that his loss of glory would be substituted by bringing sorrow and suffering to others. His overwhelming love for power did not make him flinch or hesitate in his quest to bring misery to others. His tortured and tormented mind had made a Hell out of Heaven. He would get even. He still had his faithful, re bellious angels with him but he knew fully well that he could no longer conquer by force of arms. However, he could conquer by the force of persuasion. Matchless was his strength of mind. Having lost the battle for Heaven he swore that he would now conquer Earth.

Adam and Eve being driven out of paradise.

# 7

# The Fall from Grace

**C**oiled up comfortably in his favorite tree Satan observed Adam and Eve going about their daily routine. She took care of the shrubs and plants with their beautiful and colorful flowers. He attended to the animals. Satan did not fail to notice how kind, loving, affectionate, and devoted they were to each other which irked him even more. In the afternoon when it was hot they would rest under the tree and that afforded him the opportunity to eavesdrop on their conversations.

He learned that there was a tree in the Garden of Eden known as "the Tree of Knowledge of Good and Evil" and that there was something mysterious about it.

# The Fall from Grace

God had forbidden them to eat the fruit of this tree, warn
ing them that on the day they eat from this tree they
would die. Satan, with his clear and analytical mind, was
able to discern that the fruit from that tree would not kill
them but that they would become mortal beings and
would eventually die.

This pleased Satan tremendously and he was abso
lutely overjoyed. His sly smile quickly vanished when he
learned from their conversations that there was also an
other tree and it was called *The Tree of Life*. It was right
smack in the middle of the garden and if they ate from
this one they would live forever. Satan thought that this
was an awful contradiction. One tree would make them
die and the other would make them live forever. He logi
cally concluded that Adam and Eve were neither mortal
nor immortal. They were in Eden placed precariously be
tween life and death. God had given them a free will and
they were free to choose between mortality and immortal
ity. It didn't take long for Satan to gleefully plan from
which tree Adam and Eve would choose to eat.

The Garden of Eden was a place of holiness, happi
ness, and tranquility with the grace and favor of God all
around it. Yet, God had given Adam and Eve a free will to

choose. They were free to obey Him out of love and not fear. Satan, obsessed with revenge, vowed to disrupt God's plan. He knew that he was powerless to destroy man but he would tempt him and draw him into dishonor and disgrace.

Satan observed that Eve was free to wander around alone in the garden and one day he saw her near the for bidden tree. He noticed that her curiosity made her glance at it, admiring its fruit. Surely, he told her, God was not so mean as to deny her the delicious fruit of that tree. She was utterly surprised that a serpent could talk. He, being the biggest liar, told her that he obtained the magic of speech by eating the fruit of that tree and assured her that if she ate of the fruit she too would have infinite knowl edge and be endowed with great powers, such as becom ing divine. Her better judgment warned her of danger and she was about to leave when Satan told her that she would not die if she ate of that fruit.

Her curiosity got the better of her and she contin ued conversing with the serpent. She told him that God had commanded them to refrain from eating that fruit as, He had warned, on the day they ate of it they would sure ly die. Satan told her that this was absolutely absurd and in reality God did not want them to be eternal like Him self. He employed the technique of planting a doubt in Eve's mind, thus making her a cynic by denying God's word.

Sensing the weakness in Eve he reiterated the big lie that she definitely would not die and that, as a matter of fact, she and Adam would be as great and knowledge able as God; that they would be mighty, no longer mere subjects dependent on Him.

Satan had brought disaster and ruin upon himself by aspiring to be like God, wanting all the glory and hon or from the angelic hosts in Heaven, and refusing, due to his pride, to pay homage to man, the noblest of His cre ations. He had poisoned Eve's mind by accusing God of being selfish and unwilling to share His greatness and majestic qualities with them.

Eve was confused.

She hesitated.

Satan sensed victory and hammered away with further lies and exaggerated the false hopes and the ad vantages of disobeying God's command. Eve looked up at the tree and saw that its fruit looked delicious. She saw nothing threatening about it. It was no different from the others. What harm could possibly come from it? It was pleasant to look at and it surely was a fruit to be desired since it would make her wise and eternal.

As she looked up at the tree she saw the heavens above but quickly directed her gaze back to the fruit. Af ter what seemed an eternity she reached out and plucked the fruit from the tree. Satan, with a very convincing smile, nodded his full approval.

Eve bit into the fruit.

\*\*\*

Adam realized that Eve had been away longer than usual and went looking for her. He stopped dead in his tracks when he beheld Eve by the tree with the forbidden fruit in her hand. In his anguish he felt certain that the whole Earth had felt that grievous wound. He was speechless but his mind was racing. He would beg God for forgiveness. But would God ever forgive such a bla tant act of disobedience? In her quest to gain more knowl edge, his beloved Eve had betrayed the love, the kind ness, the majesty of God. He remembered God telling them that on the day they ate the fruit of the Tree of

The Fall from Grace

Knowledge of Good and Evil they would become mortal. A sword pierced his heart for he knew that Eve would die. He loved her very much and felt that he could not live without her. Panic and despair seized him. Instead of trusting in the love, compassion, mercy, and forgiveness of God he decided he would die with her.

So he too took the fruit and ate it.

The punishment handed out by God was swift and severe. They were banished from the Garden of Eden and to prevent them from returning to eat from the Tree of Life and live forever He placed at the east of the Garden a cherubim and a flaming, revolving sword to guard the way. Heaven was now filled with sadness.

Leslie Michael

Satan returns triumphantly

# 8

# The Triumphant Return

**S**atan returned triumphantly to the sea of flames to a tumultuous reception given by the fallen angels. There were shouts of joy and wild cheering. These were followed by constant applause and standing ovations. They were welcoming a conquering hero. He narrated the events that had taken place. He assured his followers that he had only tempted Eve and at no time had he forced her to do anything. He couldn't believe God had created man, a miserable creature to start with, and then given him the freedom to choose. And even more astounding was the fact that they were weak and gullible and he now basked in the knowledge of how powerful and persuasive he was.

Standing tall and confident he told his troops that he had finally found a way in which he would get even with God in an even bigger way. He had rightly concluded that man, the miserable creature, was never satisfied with what God gave him and the pitiful example of Adam and Eve justified his way of thinking. They were given the best of everything, including a garden of delight, yet they were not satisfied. They sought to be like God and in that quest they had failed miserably and were eventually kicked out of paradise.

Satan came up with a brilliant plan. He would now start a new world order—a demonic order through the failings and weaknesses of man. Soon Satan was enjoying the fruits of his labor. He was very pleased with himself. He had a sheet spread in front of him. He rubbed his hands with glee and grinned from ear to ear every time he glanced at it. He considered himself a mighty conqueror and a meticulous bookkeeper. The balance sheet showed that for the past few millennia, millions of souls—as a matter of fact, they were as numerous as the stars and countless as the grains of sand on the seashore—had plunged into the depths of Hell. The future was very promising. For no matter how hard they tried these miser able creatures would, like autumn leaves, slowly but sure ly drift downward. Satan called for a round of high-fives with his minions. Unlike commanding officers leading men into battle, he did not have to go charging on a mag nificent steed, mowing down the enemy with his broadsword. No sir, he had it easy. All he had to do was whisper a simple but convincing lie and because of their pride, greed, lust, envy, anger, gluttony, and sloth these souls were all lost. What fools these mortals are!

He went around placidly amidst the noise and haste, whispering a lie here and whispering a lie there and the universe was unfolding according to his wicked plans. He took great pleasure and marveled at his success in so easily reducing the noblest of God's creations into pathetic, unimportant, and insignificant creatures.

God, the Father, was concerned about the fate of man. Sorrow had filled the heavens. He called for a coun cil with God the Son, and God, the Holy Spirit, along with the multitude of angelic hosts. It was clear that man had offended the majesty of God. Instead of choosing life, they had chosen death. Grace, they concluded, could no longer be extended to man without divine justice, retribution, and compassion. The wage of sin is death but God, in His infinite goodness and mercy, would give man another chance. God so loved the world that He would send His only begotten Son to die for the sins of man so that man would again have eternal life.

Mary, the virgin
*(Photo by Bernard Strothmann)*

60

# 9
# Mary, the Virgin

It seemed like another ordinary summer day. The sun had risen and like a victorious warrior it had re turned. It had conquered the night, banished all darkness, and promised the earth below another beautiful, lovely day. As it rose triumphantly into the clear blue sky the town below stirred—a little town about twenty miles west of the Sea of Galilee. It nestled on a cliff that rose more than a thousand feet above sea level. Its height and loca tion isolated it from the flow of traffic on the plain below that wound its way from Megiddo to the Sea of Galilee. It had no walls and it was unprotected. Yet, it was the si lence, the tranquility, and the serenity of the place that gave this town a very special meaning. It had a well, the only source of water, several hundred meters away from the heart of the village. Several caves in the hillside were closed in front to serve as homes. Twenty-five inches of rain fell annually and that assured its residents of good crops and pasture. There were about fifty families living there and stories abounded that the exiled tribe of Judah returning from Babylon had settled there. It was a close-knit community and everyone knew everyone.

Shepherds who had watched their flocks by night made sure that every sheep had been accounted for and began to steer them to the pastures. Farmers with sickles were heading toward their fields. Men were leading live stock to drink at the trough. Birds were chirping, telling the world how beautiful it is. Women began to trickle to ward the well to fill their pitchers with water for the needs of the day. Soon they were lowering and raising the overflowing bucket and emptying the water into their pitchers. The hustle and bustle at the well and the gossip of the women added to the bleating of the sheep and the mooing of cattle. It wasn't long before the mouth-water ing aroma of baking and cooking filled the air.

This little town of Nazareth was coming to life.

Having done her chores, a young woman sat by the windowsill. She was one of the fortunate few who lived in a modest house that was built of stone and sun-baked clay. A few shrubs adorned the front and the sides of the property. Her father would cast hopeful glances at them every time he passed that way. A vegetable patch where the family grew lentils, beans, and peas was at the rear. She watched as people hurried to and fro. Some were off to the market that was about a mile away; some to the synagogue, built in the middle of the town, to offer their prayers and tithes; and others were going to work. Vendors with trays laden with trinkets and ornaments were plying their trade. It did seem like a nice little town with happy people. Yet, all one had to do was to look deep into their eyes which betrayed their sadness. They were a conquered people.

These people were Jews occupying a land that was faithfully promised to them by God. That promise had been made to Abraham who was told that his descendants would be as countless as the grains of sand on a seashore and as numerous as the stars in the heavens. Yet, for centuries their land was invaded, conquered, plundered, and the people enslaved by pagans. Their numbers were often decimated but never wiped out. To add insult to injury they were forced to pay tribute to these hea thens.

The attractive young woman at the window was deep in thought. Her jet-black hair cascaded gently down to her shoulders. It was her eyes that held one's attention. They were soft, gentle, kind, and full of compassion— they spoke silently and eloquently of her great intelli gence. Her plain and simple dress was of a coarse materi al and she wore no jewelry.

Yet, she was a woman of grace and dignity.

As a little girl, her mother had taught her to read. She became interested and immersed herself in the study of the history of her people. She concluded that the Jews were the oldest race of people on Earth. She was aware that her people had been conquered several times and scattered throughout the world. She realized that to study the history of the Jews meant studying the history of the world.

She learned that many centuries ago her people had been captives and slaves in Egypt. She marveled at the story of Moses and particularly of his older sister, Miriam. She was a little girl barely nine years old and yet her courage would eventually lead to the end of four cen turies of bondage of the Israelites in Egypt.

Miriam had stood guarding her little baby brother who had been placed in a basket made by her mother from bulrushes and dried bark and then set adrift down the river. It had come to rest among the reeds. She was awaiting the arrival of a powerful princess, the daughter of a ruthless tyrant who had issued a decree to midwives that all Hebrew male babies were to be killed.

When the childless daughter of the Pharaoh came down to the banks of the Nile with her maidens to bathe she found little Moses in the basket. She opened it and the child wept. She was filled with compassion for this poor forlorn infant even though she instinctively knew it was a Hebrew child. It was precisely at this moment that Miri am showed extraordinary determination and fearlessness. She approached the Princess quietly and asked her if she would like her to find a Hebrew mother to nurse the baby. Without disclosing her relationship to the child Moses, Miriam brought her mother to the Pharaoh's daughter. Moses was now safe in a palace with his own mother as his nurse.

The Pharaoh's daughter had unwittingly become

part of the divine will when she found the little child. Her kindness and gentleness braved the harsh law of her tyrant father.

As the young woman at the window sill contemplated the history of her people, she was saddened by the fact that this princess had never been positively named. In biblical history she is only identified as the Pharaoh's daughter; an Egyptian pagan princess. Yet she was a woman of tenderness and compassion and she had just saved the man who would become the great lawgiver for the Hebrew nation and who would eventually lead the Israelites from slavery in Egypt to Canaan, the Promised Land.

The young woman, deep in thought, was still gazing out of the window. Her thoughts dwelled on Jochebed, the mother of Moses. It was plain to see the faith and trust Jochebed had put in her God, her Creator. She was the wife of Amram, grandson of Levi, son of Jacob. The descendants of Levi became the priestly line of the Jewish people. But as the mother of Moses, Jochebed is considered one of the immortal mothers of Israel. She lived a humble life and walked faithfully before her God. Her force and strength came from Him.

Surely, she must have instinctively known that the child Moses was destined by God for something great and she was prepared to sacrifice her own maternal love to make way for God's plan. She will always be remembered for the wise and courageous manner by which she served as a mother. For she believed with all her heart that no matter what steps the Pharaoh had taken to ensure the de struction of her people, the deliverance of the Israelites would eventually take place. She knew that she could no longer hide Moses as he reached the age of three months. During those anxious months she lived close to her God. Surely, He would not forsake her now. She had such great faith that she embarked upon a plan that was both daring and dangerous. She would place her baby in a makeshift cradle and send it floating along the edge of the Nile. The faith, the trust, and the courage of Jochebed were greatly rewarded. As the young Moses grew up, she instilled in him belief in God, their Creator, and imparted in him the divine promise God had made to Abraham and his de scendants—that they would become a great nation.

It was approaching midday. The young woman was still at the window. She saw her people go by. No people in history had ever fought so tenaciously for their liberty as the Jewish people or against such tremendous odds. The struggle of these people to regain their freedom had often seen them savagely crushed but their spirit was never broken. The young woman saw the poor, the be reaved, the oppressed, the scorned of the Earth, the old and the sick, praying, fasting, waiting, and hoping for the Redeemer who had been  promised centuries ago. They hoped and prayed with unshakeable faith that they would live to gaze upon the face of the Messiah.

The name of the young woman was Mary.

Leslie Michael

The annunciation

# 10

# The Annunciation

It was becoming unbearably hot. There was no breeze to comfort Mary in this stifling heat and not a leaf stirred in the nearby trees. The blazing noonday sun beat mercilessly down on the little dwelling as Mary rose from her seat by the window. She knew that she had whiled away the hours and felt rather guilty for not helping her mother with the household chores.

She then noticed that the curtain on the window was moving slightly, like a gentle breeze teasing a candle flame. Yet, there was no wind blowing and the stillness was overpowering. Instinctively, she knew she was not alone in the room. Turning ever so slowly she came face to face with an incredible sight. An angel in all his magnifi cent glory stood before her.

Bowing respectfully he spoke softly, delivering the most startling salutation—a poem, a message composed through all eternity and brought to her personally by one of God's most celebrated messengers, the Angel Gabriel: "Hail Mary, full of Grace. The Lord is with thee! Blessed are you among women. Rejoice, O highly favored daugh ter!"

Mary was deeply troubled by his words and won dered what his greeting meant. The angel continued, "Do not fear, Mary. You have found favor with God. You shall conceive and bear a son and give him the name, Jesus. Great will be his dignity and he will be called Son of the Most High. The Lord God will give him the throne of David. He will rule over the house of Jacob forever and his reign will be without end."

The message of the angel was truly disturbing. As the message began sinking slowly into her consciousness, Mary's keen and analytical mind was working feverishly. She did the unthinkable. She challenged the angel! "How can this be since I am a virgin and do not know man?" The angel answered her, "The Holy Spirit will come upon you and the power of the Most High will overshadow you. Hence, the holy offspring to be born will be called Son of God." And to emphasize the power of God, the an gel continued, "Know that Elizabeth, your kinswoman, has conceived a son in her old age. She who was thought to be sterile is now in her sixth month. For nothing is im possible with God!" Bowing humbly and obediently, Mary said, "I am the servant of the Lord. Be it done unto me according to thy word."

The angel departed and a great transformation took place. Mary was now truly a woman with child. Not just any woman and not just any child. She had been cho sen to be the Mother of the Messiah. She bounded with joy from her home to see Joseph, a good and upright man to whom she had been recently betrothed. She must tell him. He would also be happy and surely he would under stand what was happening to her.

Joseph, the betrothed

# 11

# Joseph, the Betrothed

Joseph, of course, did not understand anything. He was in a state of total shock. How could something so beautiful go so terribly wrong? He had always loved Mary. He had watched her blossom from a little girl to a truly beautiful woman. She was now fifteen. She was in telligent, graceful, and always held herself with dignity. And those eyes! It was maddening not to see her except once a week at the synagogue. None of the other women had interested him. He was now twenty-one and every one knew that he had not yet been betrothed because he had been waiting for Mary. He was hesitant to ask for the hand of this lovely maiden because he came from a poor

family. On several occasions he had come by on the pre
text of running errands and would linger with Mary, leav
ing Anna, Mary's mother, in a state of dismay and ill-fore
boding.

Mary was so beautiful that her parents had always
entertained the notion there would be numerous other
suitors; perhaps some of them even wealthy. Anna had
guarded her daughter with all her strength and had even
made it very clear to Joseph and his parents that her love
ly daughter was destined for the finer things in life, and
not meant to be just another housewife slaving away and
having children for some poor ordinary carpenter.

Yes, there had been other suitors as well. But there
was something disturbing about the whole situation. Like
Anna, the Jewish people awaited a Messiah. It was a na
tional obsession. In Hebrew, the word is Messiah—the
anointed. Jews waited breathlessly for his coming. No
matter how many generations went by each new genera
tion was sure that theirs would be the one to greet him, to
welcome him as a mighty king accompanied by legions of
angels that would crush Israel's enemies, reign over the
pagans, and bring them to the one true God, YAHWEH.
All the prophets were in agreement that the Savior would

be from the House of David. He would be called the Son of Man. There would be a time of peace, love, forgiveness, and happiness. It was a pleasing and wonderful image for the Jews except that they did not know when or where the Messiah would come and how they would recognize him. However, every Jew knew the prophecy of Isaiah—that a virgin shall be with child and bear a son and shall name him Immanuel.

Like Mary, Joseph was also from the house of David and could trace his ancestors not only past David but to Abraham as well. This identified Joseph with Israel and further down, his ancestors could be traced to Enoch, son of Seth, who was the son of Adam who was the Son of God.

The descendants of David now numbered in the thousands. By the age of fifteen, women had been betrothed and as these virgins, along with their grooms, were led to the bridal chambers they prayed and hoped that they would be the one who would bring forth the Messiah. Anna, the mother of Mary, contemplated all this. She wanted the best for her beautiful daughter. Mary was not only lovely but very spiritual. She knew that Mary would not be happy with the others who asked for her hand in marriage. Joseph was, after all, a nice young man. He was tall and handsome. He was patient and kind. He had a winning smile and was always helpful to his neighbors. On the Sabbath, whenever he was requested to read the Holy Scriptures, Anna sensed the deep spirituality in him.

Joseph had learned his trade from his father who earned his living by fixing a broken yoke, mending a trough, and other odd jobs. Joseph was a good worker and soon he was building furniture and houses, as well. He dreamed of building a beautiful home for Mary, whom he absolutely adored. To the other women in the town it was obvious he only had eyes for her. They were also sure that if he did not marry Mary he would not marry at all, and that would be a terrible disgrace and a big disappointment to his family.

Joseph finally made a decision. He approached his father, Jacob, and implored him to forget all his fears and approach Joachim on his behalf for the hand of Mary. Jacob hedged and delayed for several days until one day he bathed and put on some clean clothes and left for Joachim and Anna's place. Timna, the wife of Jacob, was nervous and fearful that Mary's parents would laugh at her hus band because they really had nothing much to offer and was apprehensive that they would treat Jacob with con tempt because they were very poor.

Joseph wondered if he had made a mistake. He loved Mary and had waited so long for her. Not as long as Jacob, his ancestor, had waited for Rachel, but it seemed just as long. He resigned himself to the possibility that he would be rejected and that that would result in the loss of Mary forever. He had just about given up all hope when his father came joyfully down the street and almost crashed through the door. He told them that he had a long talk with Joachim and Anna. It was now clear to Anna that it was really her pride and not Mary's happiness that consumed her thoughts. Joseph was a good and righteous man and would make Mary happy. Joseph had been ac cepted.

Anna got up slowly from the couch. There was much to be done if the betrothal was to be announced on the coming Sabbath and if guests were to be received and entertained after the ceremony. The terms of the contract of marriage would be worked out later. Now was the time for rejoicing. Wine flowed freely and there was a lot of singing. It was not long before the whole town knew what had transpired.

It had been another glorious summer day. The sun had shone brightly as people entered the synagogue. The menorah with candles in its seven branches were like tongues of flame. The readings from the Torah had been chosen. Mary and Joseph were summoned and the solemn service began. At the end, Joseph, son of Jacob, had requested the hand of Mary, daughter of Joachim, and had been accepted. It was agreed that the wedding would take place within a year, giving time for all con cerned to make preparations. Until such time the couple would continue to live with their parents. But as of that day, in the eyes of the world and spiritually before God and man, Joseph and Mary were husband and wife—but not in the flesh.

The plight of Joseph

# 12

# The Plight of Joseph

**I**t had hardly been three months since the betrothal and now Joseph's life was turned upside down. He could scarcely believe Mary's words. "I am with child!" That is what she had told him. It was as if some one had struck him a heavy blow. How could this possibly be? He was an honorable man. He would never bring shame and disgrace upon himself, his beloved Mary or her family. Then whose child was it?

Even as he waited for the answer he was frozen with fear and dread. She had betrayed him. The one word that he refused to utter came horribly to his mind. Adultery! He also knew the penalty for that sin. Stoning! Even though he loved her with all his heart, even though she undoubtedly had been unfaithful, he could not bear to see her face that harsh penalty. She had explained to Joseph that the child in her womb was the Son of YAHWEH. She narrated all that the angel had told her. She emphasized that she had been truly faithful to him. But God had now chosen her to be the mother of the Savior. Joseph could not accept this. How could God have chosen to have this honor bestowed upon Mary, a woman already spoken for in marriage to a man who loved her ever so dearly? That was so unbecoming of God.

He was beside himself with righteous anger. He just could not understand it. He calmed himself and began to think a little more clearly. Yes, he would marry Mary. He was, after all, betrothed to her. He was an upright man and would do the honorable thing and would not expose her to the harsh realities of the law. That marriage would save her from the stoning for adultery. He would then divorce her quietly.

Such were his intentions and in a fitful sleep that night, Meteron, an angel of the Lord, appeared to him and said, "Joseph, have no fear in taking Mary as your wife. It is by the Holy Spirit that she has conceived this child. She is to have a son and you are to name him Jesus because he will save his people from their sins." Joseph then realized that all this had happened to fulfill what the Lord had said through the prophet: "The virgin shall be with child and give birth to a son and they shall call him Immanuel."

Meanwhile, Mary's thoughts now turned to her kinswoman, Elizabeth. She remembered what the Angel had said and made preparations to visit her. She lived in a little town called Ain Karem, a few miles from Jerusalem. Such a journey would probably take four days. It would be a perilous undertaking. Mary refused to consider all the problems she would have to face. It was the urgency and the importance of the journey that helped her over come the problems and the dangers.

Elizabeth's husband was Zechariah, a priest. Both of them were just in the eyes of God, blamelessly follow ing all the commandments and ordinances of the Lord. They, however, were childless and Elizabeth was well past her childbearing years. She was an upright woman who walked faithfully before her God. She spent most of her time in prayer and was sure that if God could create the heavens and all the countless stars that shone brilliantly he would answer her prayer to bring forth a child.

One day it was Zechariah's turn to fulfill his func
tions as a priest before God and the full assembly was
praying outside. An angel of the Lord appeared to him,
standing at the altar. Zechariah was deeply disturbed and
overcome by fear. The angel said to him, "Do not be
afraid. Your prayer has been heard. Your wife, Elizabeth
shall bear a son whom you shall call John. Joy and glad
ness will be yours at his birth, for he will be great in the
eyes of the Lord. He will never drink wine or strong drink
and he will be filled with the Holy Spirit. Many of the
sons of Israel will be brought back to the Lord, their God.
God himself will go before him in the spirit and power of
Elijah, to turn the hearts of fathers to their children, and
the rebellious to the wisdom of the just, and to prepare for
the Lord a people well-disposed."

Zechariah did not comprehend this and said to the
angel, "How am I to know this? I am an old man. My
wife, too, is advanced in age." The angel answered, "I am
Gabriel who stands in attendance before God. I was sent
to speak to you and bring you the good news. But now
you will be mute—unable to speak—until the day these
things take place, because you have not trusted my
words. They will come true in due season."

Meanwhile, the people were waiting for Zechariah, wondering at his delay inside. When he finally came out, he was unable to speak and they realized he had seen a vision. He kept making signs to them for he remained speechless.

Elizabeth was in the sixth month with child when Mary arrived at her home. When she heard Mary's greeting, the baby leapt in her womb. Elizabeth was filled with the Holy Spirit and cried out in a loud voice, "Blessed are you among women and blessed is the fruit of thy womb." And in a humble voice she asked, "But who am I that the mother of my Lord should come to me? The moment your greeting sounded in my ears, the baby leapt in my womb for joy. Blessed is she who trusted that the Lord's words to her would be fulfilled."

On hearing those words Mary was filled with understanding and gratitude. In a humble yet prayerful tone she proclaimed that her spirit had found joy in God, her Savior. Her very soul magnified the Lord who, in spite of her lowliness, had looked upon her with such favor that, henceforth, all generations would call her blessed. God who is so mighty had done great things to her and holy will be His name. This wondrous hymn of praise for

God's greatness poured from her heart and she pondered the mighty things that God had done to her. Mary could hardly contain her happiness. She was now the daughter of the Father, bride of the Holy Spirit, and the mother of the Son.

Mary remained with her kinswoman for three months. When Elizabeth's time for delivery arrived she gave birth to a son and there was great rejoicing among her neighbors and relatives as they assembled for the cir cumcision. They intended to name him after his father when the mother intervened, saying that his name will be John. They reminded her that none of her relatives had that name. Then, using signs, they asked the father what he wished his son to be called. Using a tablet he wrote the words to indicate that his son's name would be John. At that moment his mouth was opened and his tongue was loosed and, like Mary, he began to speak in praise of God.

Mary returned to Nazareth. Surely there would be social problems regarding her forthcoming marriage to Joseph and the serious implications of her now obvious pregnancy. She placed all her trust in her God. She would remain faithful to her Creator and was confident that He would see her through this.

Her words, "I am with child," still haunted Joseph. All his life he had waited for her. It was such a betrayal. He knew that his parents were also disappointed and heart-broken. He sensed that they wished he hadn't wait ed so long for this one maiden. He could have easily mar ried some other woman and by now his parents would have had grandchildren. But he had loved Mary with all his heart and now he was angry, confused, and agitated. At the same time he did not want to bring shame and dis grace to her and to her family. He remembered the words of the angel telling him to fear not and to take Mary as his wife for the child in her womb was conceived of the Holy Spirit. Recalling the words of the angel calmed him down.

He also recalled the line from the Psalms: *"Be still. Know that I am your God."* He knew then that he would do the just and honorable thing. A few weeks later Mary and Joseph were married in a quiet ceremony.

Caesar Augustus

# 13
## Caesar Augustus

**C**aesar Augustus was very pleased with himself. Sitting in his palace in Rome, sipping wine from a goblet, he looked with great satisfaction at a map that was spread in front of him. He was the first Emperor of the great Ro man Empire and had been its ruler since 27 BC. In spite of the usual conflicts on some distant frontier there had been a period of relative peace for some time. The Empire had spread far and wide. Much of the Mediterranean coun tries had been conquered. Not long after, Judea had been added to the province of Syria. The Romans were great warriors and they built the finest roads which enabled them to move men and chariots swiftly and effectively. Every country that fell to them saw roads being built with the labor of the conquered people. They were then taxed to help Rome pay for the roads. Caesar Augustus prided himself with the fact that all the conquered nations were now subject to Rome. There was only one capital and that was Rome and only one leader and that was him. Fully aware that the city was architecturally unworthy of its po sition as the capital of the great Roman Empire, Augustus constantly sought to improve her image. He boasted that he had found Rome built with bricks and he would leave it clothed in marble.

Augustus was born at dawn on the 23$^{rd}$ of Septem ber, 63 BC. As a young warrior he adopted the surname Caesar. A few years later the title Augustus was added. Some of the senators wished him to be called Romulus, after the first founder of that great city. He argued that the title "Augustus" was more original, honorable, and exalt ed. He convinced the senators that sanctuaries and other notable places in Rome that were consecrated by their holy priests were often referred to as "august" implying that it gave those places an increase in dignity.

In battle, Caesar Augustus not only displayed skill

as a commander but courage as a soldier. His officers were fully aware that Augustus condemned haste and recklessness. He held high a principle that no battle should be fought unless the hope of victory was greater than the fear of defeat. But he was also ruthless. The be sieged city of Alexandria, to where Marc Anthony and Cleopatra had fled, he razed to the ground. Anthony sued for peace but Augustus forced him to commit suicide and he was so eager to claim Cleopatra as an ornament for his great triumph that he summoned snake charmers to suck the poison from her from the bite of an asp.

The official calendar reformed by Julius Caesar had been neglected. He set about correcting it and in the process renamed the month of Sextilis, "August," after himself, even though he was born in September. All Kings and rulers who were allied and who enjoyed the friend ship of Augustus were quietly pressured to found a city in their respective dominions and name it "Caesarea." Au gustus was remarkably handsome and his eyes were clear and bright and he fancied that they shone with a divine radiance. It gave him immense pleasure if anyone whom he glanced at would lower his head as though dazzled by the brilliant rays of the sun.

## Caesar Augustus

Caesar Augustus was not only a great warrior and a great Emperor he was also a great bookkeeper. He decided to issue a decree to the whole world that required every Roman subject to be registered in his own city. He wanted to make absolutely certain that every one of his subjects, even those on the outermost fringes of the Empire, paid their taxes, even when he was told that the Jews were so poor they had only one god.

And so it came to pass that in the little town of Nazareth, Roman soldiers put up notices everywhere informing the citizens that all Roman subjects had to be registered in the town of their origin. The notices were written in Latin, Greek, and Aramaic.

Leslie Michael

Satan and his minions

# 14
## Satan and his Minions

$S$atan had been enjoying the fruits of his labors. He was very pleased with himself. He had a sheet spread in front of him. He rubbed his hands with glee and grinned from ear to ear every time he glanced at it. Like Caesar Augustus, he too was a good bookkeeper. The bal ance sheet in front of him showed that for centuries, mil lions of souls, countless indeed, as the stars in heaven had plunged into the Sea of Flames. Satan always appeared to those inferior creatures as a good and benevolent angel ever ready to guide them in their trials and tribulations. Ever ready to help them acquire more and more earthly possessions He charmed and convinced them that wealth brought power and glory—so much glory that would cause men to look upon them with great admiration and be desired by so many beautiful women. And as a gra cious host, Satan always warmly welcomed them to his fiery abode.

Satan loved people like Caesar Augustus. They were good for his business. People like him were not sat isfied with being strong and handsome but wanted to firmly believe that they also possessed eyes that shone like the brilliant rays of sunshine. They basked in pride and vain glory, desiring their capital cities to be magnifi cent monuments to their perceived greatness. They just had to be larger than life. Their egos were incredibly in flated and they wanted to be remembered, not for their love, kindness, and mercy to their fellow human beings, but by demanding that cities in other countries be named after them. They were brutal, violent, and bloodthirsty. Not satisfied with what they had, they marched their le gions into other kingdoms butchering, ravishing their women, enslaving them, taking their lands, their posses sions, and crushing them under the conqueror's heel, leaving their soil soaked with the blood of innocent peo ple.

What wonderful friends Satan had in people like Caesar Augustus. Hail Caesar! Way to go! What a shining and august example he was showing. With "Welcome" and "No Reservations Required" signs prominently displayed, the sea of flames was getting overcrowded. The fear, the uncertainty, the hopelessness, the shock of being in that wretched place, coupled with the oppressive heat and utter loneliness had begun to take their toll and the inmates were wailing and mourning. Their tortured minds began silently screaming and begging for mercy. Satan had that evil look and smile on his face. He grinned and viciously reminded them that Hell, after all, was not meant to be a comfort zone.

Satan just couldn't believe his luck. His wicked plans for conquering the Earth were far easier than he had anticipated. He gleefully imagined how glorious that day would be when he crowned himself commander-in-chief of the world. His thoughts turned scornfully to Archangel Michael and wondered what he would do now to stop him. Yes, Michael and his loyal angels had fought brilliantly and valiantly and had emerged victorious. He wished and hoped that Michael would embark on another campaign to engage him in the battle for Earth. Satan, af

Satan and his Minions

ter all, still had his rebellious angels with him and he rea soned that all Michael would have would be those pathet ic human beings.

Leslie Michael

The birth of Jesus

# 15
## The Birth of Jesus

J oseph was not too pleased when he read the no
tices that had been put up by order of Caesar Augustus. It
was a long way from Nazareth in Galilee to Bethlehem in
Judea. He was, by that decree, bound to go there. His par
ents had lived in Bethlehem and he was born in that city
of David. However, Mary's time to give birth was not too
far away and she, being his wife, was obliged to go with
him. Defiance of that law would mean harsh penalties,
perhaps even imprisonment. Joseph agonized over the sit
uation and Mary's parents were beside themselves with
worry and anxiety. It was Mary who reminded them that

her place was with her husband, come what may, and that she would accompany him to Bethlehem.

As Joseph prepared for the trip, it occurred to him that he would have to go through Samaria. This caused him grave concern as there was some hostility between the Samaritans and the Israelites. The Samaritans were a community of Jews who claimed to be blood relatives of those Jews of ancient Samaria not exiled by the Assyrians. Jews who returned to their homeland after the Babylonian exile would not permit the Samaritans to build the Second Temple in Jerusalem. This prompted the Samaritans to build their own temple in Shechem at the base of Mount Gerizem. Consequently, the enmity and the contempt that both held for each other convinced Joseph that he would have to take a much longer route.

With Mary on their trusted little donkey they trav eled in an easterly direction and then across the plains of Esdraelon to the Jordon River and then southward toward Jerusalem. He was glad that it was spring and some of the mountainous passages were easy to cross. He thought of the other conquered nations in Europe that lay in the Alps and of the people, who were also subject to the decree of Caesar Augustus, trekking across treacherous snow cov ered mountain passes to towns of their origins. It was a hard, long, tedious journey and Mary and Joseph arrived in Bethlehem completely exhausted.

Joseph was optimistic that he would, as he entered the city of his ancestors, find some accommodation for Mary, especially in view of her condition. In vain he went from one house to another only to be turned away. Deject ed, he climbed a hill to where a faint light was burning. That must be the inn! He remembered. As a little boy he played games on this hill with his friends He would sure ly find lodgings for Mary and himself here. At the Inn, there were rooms for the Roman soldiers sent to keep or der and ensure everyone was properly registered for tax purposes. There were rooms for the rich merchants dressed in fine garments. There was room for anyone who could discreetly slip a coin or two into the greedy hands of the innkeeper. But there was no room for Mary and Joseph. It was close to midnight.

Going around the hillside he remembered there had been a cave. Once inside the cave he lit a lamp, gath ered some straw, set Mary gently down and made her as comfortable as possible. There, in the shelter of the cave, in the still of the night and away from all the revelry, drunkenness, and debauchery at the inn, Mary gave birth to her Son. Jewish law conferred preferential status on the eldest son even if he were the only child. This child was considered holy. He belonged to God. Mary smiled at her little baby, wrapped her first born in swaddling clothes, and laid Him in the manger. The redemption of mankind had begun.

Satan was livid with rage. He had convincingly outsmarted God the Father. Now he would have to deal with the Son. Satan was never one to back away from a fight. He was not one to turn tail and bolt. He was still reeling from this totally unexpected blow—the birth of the Savior—and knew that somehow he had to come up with a plan. His whole wide and wicked world had been turned upside down but he was confident that he would stand up to the challenge. Surely a helpless baby and his mother could not possibly pose as a serious threat. It was an inconvenience, perhaps—but definitely nothing to be unduly concerned about.

He was still thinking about the baby when his thoughts turned to dwell on the baby's mother, Mary. Yes, she was a woman but so different from Eve. He had never seen any woman like her before. She had a quiet and serene beauty about her. She was silently strong, fierce in her love for God, implicitly obedient to His will. It seemed that throughout the ages she had been the one chosen to be the mother of the Son of God. She was pure, holy, spotless, and immaculate. She would nurse the child, protect him, care for him, love him, and above all, set him firmly on his divine mission. It was also foretold that this woman would be the one to crush the head of the serpent who had seduced Eve.

The more Satan thought of Mary the more he hated her. But he knew that hating this woman was not going to get him anywhere. He had to think and think hard and come up with a plan to destroy the mother and child. He was deep in thought when he noticed three men glancing at the heavens. They seemed to be astrologers and they were studying the stars. They seemed to be looking for something. Satan never slept and nothing escaped his attention. He resolved to find out who those men were.

Solomon and Sheba

# 16

## Solomon and Sheba

**K**ing Solomon was sitting in judgment. The case before him was rather unusual. Two women were before him in order to settle an issue as to which one was the true mother of a beautiful little baby. When Solomon as cended the throne after his father, King David, he had prayed to God for an understanding heart to judge his people. Because he had not asked for wealth nor sought the death of his enemies God granted him that wisdom. He now silently prayed that God would help him to be discerning in administering justice in this case.

He summoned a guard and commanded him to kill

the child and divide it in half and give one half to each of the two women. One of the women fell at his feet and begged him to spare the child saying that she was pre pared to give the child up to the other woman rather than see the child killed. King Solomon was now able to dis cern who the true mother was and handed the baby over to the woman who showed such love and compassion for the child.

The fame of Solomon's wisdom spread far and wide. He was the son of King David and Bathsheba. Even though he was not the first-born son of David, his father had promised Bathsheba that Solomon would be the next king and while he was still alive he commanded his ser vants to have Solomon brought to the Gihon Spring and had a priest anoint him. When Solomon ascended the throne he had inherited a vast empire that extended from the Euphrates River in the north to Egypt in the south. The building of the Holy Temple in Jerusalem was his crowning achievement. It took more than seven years to be completed and it was built of stone and cedar and overlaid with pure gold.

During his long reign, King Solomon's kingdom achieved its highest splendor. He surrounded himself with all the luxuries and the external grandeur of that age, acquiring more than 10,000 horses and 1,000 chariots. Is rael also enjoyed great commercial prosperity due to ex tensive traffic carried by land and sea to distant countries. For centuries southern Arabia had been a trading center for gold, frankincense, myrrh, and various spices. Much of the wealth of King Solomon came from the traffic of the spice merchants. Other monarchs and rulers sought his advice and to strengthen their alliances.

Makeda, the woman who ruled the ancient king dom of Sheba, heard of the great wisdom of King Solomon. She was one of the many warrior queens who ruled Nubia which lay south of Ancient Egypt and through which the Nile River flowed. Nubia was bor dered by the Red Sea.

Ancient Egyptian history tells of trade missions in Nubia where Egyptian merchants imported gold, incense, and ivory, thereby increasing the prosperity and stability of both countries. Nubia, now known as Ethiopia, was eventually divided into smaller kingdoms, including She ba. The queen journeyed from Axum, the heart of Ethiopi an art and culture, to Jerusalem. She wanted to meet King Solomon and seek his advice on matters of government.

She was very wealthy and set out bearing gifts of gold, precious stones, spices, and other treasures for the Temple that Solomon had just finished. The Temple would one day house the Ark of the Covenant. She was impressed by his wisdom and gave her blessings to Solomon's God.

They spent much time together discussing their two great kingdoms. When the Queen returned to Axum she gave birth to a son who eventually traveled to Jerusalem to meet his father. King Solomon was overjoyed and advised his son, Menyelek, to stay and study the ways of the Jewish people.

King Solomon loved many foreign women even though the Lord cautioned him that they would turn his heart to their gods. But Solomon fell in love with them. He had 700 wives and 300 concubines building idols for their gods. Those lavish expenditures exhausted the country's treasury and even though it was vast it would be found wanting as he continued maintaining the pride and vanity of all these women. He began to neglect the business of governing his empire and by attempting to keep up the grandeur the burden soon fell upon his subjects and they were levied with more taxes.

There was a time when God had loved Solomon

and delighted in him. But now he was angry with him. Solomon had turned away from God, committing willful disobedience and doing the very thing that God had for bidden—worshipping other gods. The Lord was very dis pleased and raised adversaries against Solomon. Half of his kingdom revolted and soon his kingdom was threat ened. Menyelek and other pious Jews decided that King Solomon was unworthy to be the custodian of the Ark of the Covenant wherein lay the presence of the Lord and so they secretly plotted to cunningly whisk it away. Nubia had stretched forth her hand to God and to that country went the Ark of the Covenant. Temple priests and their families accompanied it and stayed in Nubia to teach the people the ways of the Lord. Other Jews followed after the destruction of the Temple.

Solomon, the wise guy

# 17
# Solomon, the Wise Guy

**S**atan could not help but sneer at King Solomon. Here was a man famed for his wisdom and yet acting like a downright fool. Satan himself had a difficult time figuring out how Solomon had ascended the throne in the first place. He certainly was not the eldest son of his father but the fourth son of King David's seventh wife. Satan chuckled. Wow, this was far better than he ever imagined. Those miserable human beings—those mortal fools knew how to cut their own throats. When Solomon eventually became king he had prayed to the Lord to grant him a discerning mind. Because he had not asked for wealth or the

destruction of his enemies, he had been granted that wis
dom. Satan wondered who the bigger fool was—God or
Solomon? Did not God understand that Solomon was
powerful and his mighty army could wipe his enemies
out? He did not need God's help with that. Or that
Solomon could, without hesitation, very easily resort to
the game of political intrigue, back stabbing, blackmail,
treachery, betrayal, and murder to get rid of his adver
saries? After all, that was the order of the day—absolute
child's play.

He did not ask God for wealth because he already
had it. His marriage to the Pharaoh's daughter brought
peace and prosperity between their two countries and, as
part of his dowry, the Pharaoh had gifted to Solomon the
Canaanite fortress of Gezer. The treaty with Hiram, the
ruler of Phoenicia, enabled Israel to be absorbed into the
oriental trade and commerce routes bringing great wealth
to Israel.

It seemed that Solomon, with all his power and
wealth, really had nothing to show for it. As far as morals
and integrity were concerned, his father, King David, was
no role model either. As a matter of fact, David's hands
were stained with the blood of innocent people. David's

first wife had died without bearing him any children. However, she had five sons from a previous marriage. David handed them over to the Gibeonites to be killed to avenge their grandfather.

One night while David was restlessly pacing the roof of his palace he observed a beautiful woman bathing. She was Bathsheba and was the wife of one of his gener als, Uriah, who was away at war. David summoned her and spent the night with her. When she became pregnant David sent for Uriah so that he may lie with his wife and, therefore, have the true paternity of the child concealed. When Uriah refused he was ordered back to the front lines where he was killed in battle. David and Bathsheba were then married.

Satan was truly enjoying the soap opera. Here was a wealthy man who had many sheep, yet he stole the one beloved sheep of a poor man.

King David's son, Solomon, also had a weakness for women and he loved strange women; the stranger the better, even if they were pagans that God had expressly forbidden saying to the children of Israel that they shouldn't go into them and neither should they come in unto them because they would turn their hearts away from the Lord.

Satan was tremendously pleased when Solomon starting loving those women and in the process had collected about 700 trophy wives and 300 concubines for good measure. Solomon put on a dazzling display of his power and wealth by bestowing upon them expensive gifts, most of them more ornamental than useful, and throwing lavish, drunken, and immoral parties for his trophy wives. Fully inflamed and defiant, he considered himself modern and emancipated. With insatiable lust he assured himself of his own virility. He also erected monuments and idols for his pagan women, burning incense and offering sacrifices to their gods.

Satan waited patiently. As someone gifted and famed for his wisdom, Solomon was a miserable failure and cut a sorry figure among his own people. He was morally bankrupt and he had exhausted himself and the treasury. Satan knew fully well that all good things have to come to an end and it was not long before God gave up on Solomon.

A Nubian king

# 18

# The Kingdom of Ethiopia

Solomon's kingdom was inherited by his son, Re hoboam. He was the son of one of the wisest men in histo ry yet he did not inherit his father's wisdom. He was im prudent in the affairs of government and made foolish de cisions and unwise treaties. Fearing a civil war between Solomon's sons, more Jews left Israel to resettle in Ethiopia. A once strong and unified nation was now weak and divided and eventually split into two. It was not long before the Assyrians conquered the land and the Jewish people were sent into exile and were scattered.

Meanwhile, the kingdom of Nubia flourished. The

succession of kings there claimed to be descendants of King Solomon and since Solomon was an ancestor they were also kinsmen of the Lord. They maintained a strict biblical code of purity concerning life, death, ritual clean liness, and dietary laws. They safeguarded the Holy Books and Torah scrolls.

Prior to its acceptance of the Jewish faith the king dom of Nubia was a nation of star-worshippers who stud ied the sun and the moon. They closely observed celestial objects and charted their movements. They perfected the system of interpreting any phenomena occurring in the heavens. The rise of a new moon in a cloudy sky would mean a great victory on the battlefield. The orderly move ments of heavenly bodies were recognized and studied.

Ancient astrologers were known as inspectors of the heavens. Their lives and happiness were guided by the movements of the celestial objects. They were able to study the various locations of certain stars, analyzing their distance and brightness. The connections between the cosmos and the Earth below played a significant part in their lives. It was important how those movements were interpreted because to them the heavenly bodies were God's creations, with definite roles to play in the universe, but subordinate to their Creator. The emphasis was always on YAHWEH, their Creator. Their study of ancient Hebrew texts revealed that King Solomon himself had said that God had given him a sound knowledge of existing things—that he might know the organization of the universe and the force of their elements. The position of the stars and the power of the winds were also revealed to him by God. The Nubian Jews studied the heavens without worshipping the celestial objects because they be lieved worshipping them would be akin to magic and sor cery, and thus amounting to blasphemy. They scanned the heavens and looked for signs that would herald the birth of the Messiah, for it had been foretold that a virgin would bring forth the anointed one. They longed for the kingdom of the Messiah and Israel's past glory and prayed for the everlasting kindness of the Lord, despite their ingratitude and infidelity.

The three astrologers

# 19

## The Three Wise Men

**S**atan was still reeling from the fact that the sav

ior of mankind had been born. He had been so sure every

thing was under control. Now he seemed utterly helpless.

It was then that he observed the three astrologers gazing

at the heavens and noted that they were well-read and ed

ucated. Like other races and cultures they paid great at

tention to the stars and other heavenly bodies and be

lieved they could foretell the future by studying the

movement of celestial objects.

At the time when the three astrologers were scan

ning the heavens there was a period of unparalleled peace

and prosperity in the Roman Empire. Gala festivities were being held throughout the kingdom and in other prov inces that were allied with Rome. The most significant event that occurred in this period was the bestowing upon Caesar Augustus the title of *"Pater Patriae"* or "Father of the Empire." This was the highest honor any Roman could achieve.

Between 7 and 2 BC, there were several astronomi cal events. During an eighteen – month period between 3 and 2 BC, people witnessed some of the most spectacular displays in the  heavens. These extraordinary celestial events inspired many wonderful and mystical interpreta tions by priests and religious people. They had taken place when the Roman Empire was in joyful celebration. It seemed as though the greatness of Rome and, in partic ular, the great Caesar Augustus, were being confirmed by the heavens.

Except for the Jews, there was no clear distinction between jugglery, magic, astrology, and astronomy. The movements of the celestial bodies were observed and charted and the information was used to determine the course of history. And that information was believed to be reliable and scientific. Most people, especially rulers of various countries, were firm believers in these events as strong indicators of the present and the future.

Satan was still wondering what to do as he observed the three astrologers. His keen and analytical mind reminded him that he was once the brightest star before he fell from grace. He was, after all, the magnificent Lucifer, created from fire. He was the "Bearer of Light," the "Shining Star of Dawn," and, "Son of the Morning." Now he was called the "Prince of Darkness." This was down right insulting. He would show them.

A brilliant plan slowly and surely formed in his mind. Satan could appear in any form and was the master of disguises and lies. He usually appeared as a kind, gentle, helpful, and benevolent person. He could fool anyone as he was also very powerful. He could hardly believe that God had given him such power and man only a free will. He recalled his encounter many centuries ago with a certain man named Job. With the power given him he had virtually destroyed Job and everything he possessed. Even Job could not help but admit that Satan's light was brighter than the noonday sun. He was definitely associated with light and the sky. He would now use that power and his unimpeded intellectual energy.

The three astrologers, still looking at the skies, suddenly noticed something strange. A star! It was unusually bright, extraordinary and magnificent, and they observed it at its rising. It was now dawn. It hadn't been there before and now it was moving. They were thrilled beyond words as they had never seen anything like this before. They had discovered something. They knew and believed like most people that stars foretold significant events. Something had happened or was about to happen. They would be the first to tell the world about it. In their excitement they failed to realize that there is absolute law and order in the universe. This star was moving erratically and in the wrong direction. They nevertheless followed it and it led them to Jerusalem. Dressed in their finest clothes, they presented themselves to King Herod the Great.

Leslie Michael

Slaughter of the innocents

# 20
## King Herod

**K**ing Herod was from Udemea. He was a brave warrior and a man of great physical courage and skill; a perfect marksman and a mighty hunter. He fought along side the Romans who slew more than 12,000 Jews and witnessed another 30,000 sold into slavery. After many conquests by the Romans, Judea was added to the Empire by joining it to the Roman province of Syria. When An tipater died, his son, Herod, was named King of Judea. This Udemeanian, with the help of Roman funds, ruled Judea. During the time of peace that prevailed, the Roman Empire prospered along with its conquered territories.

# King Herod

Herod was an intelligent man but had no morals, scruples or honor. He was not a Jew by religion or conviction but many said that he was half Jewish. He slew all the Jewish leaders and became one of the most colorful rulers in history.

He detested the ugly synagogues of the Jews. He was more impressed with Greek culture and encouraged Greek ideas in dress, literature, and art. He built, at great cost, a theatre and amphitheatre in Jerusalem and adorned them with monuments to Augustus and to other heathen gods. He shocked everyone by tearing down the 500 year-old temple and built his own great temple on that site. He dreamed of leaving Judea more beautiful than it had been before and, hence, he was given the name "Herod the Great." Ironically, he was also referred to as "King of the Jews."

He ruled by terror. Any uprising was ruthlessly crushed. He had his opponents disarmed and backed into a narrow space so that his soldiers, in full battledress and armed with shields and swords, could slaughter them. They were never killed by the simple thrust of a sword. The soldiers always obeyed their orders and hacked them to pieces. Herod was filled with hatred. He was also inse cure and murdered anyone who posed as a threat—real or imagined. He was despised for his ruthlessness and his alien descent. He became king by his guile and under the protection of Rome. The Romans tolerated him as long he kept the peace and the people paid their taxes.

He married a woman named Mariamne and she bore him two sons. She was a princess of the Maccabean line and by this marriage and through her royal blood he gained an additional claim to the Jewish throne. In spite of the peace and prosperity of that time, Herod, degener ated miserably into the depths of suspicion, insecurity, anger, ruthlessness, and insanity. He accused Aristobolus, the brother of Mariamne, of plotting against him. Aristo bolus was a dashing young man and well-liked by the Jews. The Jews found hope in him and saw in Aristobolus the prospect of the Maccabean line coming again into power. Herod murdered the young Aristobolus and this caused him to sink further into disillusionment. He now accused his wife of conspiring against him and assured his trusted advisers that he had absolute and concrete evi dence that such was the case. It was not long before Herod gave the order. His soldiers marched into Mari amne's bedroom and hacked her to pieces.

# King Herod

With the brutal killing of his innocent wife, Herod became a raving lunatic. He could be seen and heard screaming in his palace. To pacify himself he married a succession of women and had several children. Over time, other suspicions and plots developed in his fevered brain. Under terrible torture, the accused confessed to their crimes and implicated other innocent people. They, too, were slaughtered.

Sinking deeper into madness, he turned against his own children, the sons of Mariamne, and accused them of plotting to overthrow him. He was advised against harm ing them for they were popular and were liked by every one. Soon Alexander and Aristobolus (named after his un cle) were murdered along with anyone who held them in high esteem. Under the reign of King Herod, tens of thou sands of Jews and others were also murdered.

Leslie Michael

The star of Bethlehem

# 21
# The Star of Bethlehem

It was into the palace and court of this raving and evil king that the three astrologers came. They were dressed like royalty and carried themselves with an air of sophistication and wisdom. They inquired, "Where is the newborn King of the Jews? We observed His star at rising and have come to pay Him homage." Herod, on hearing this became greatly disturbed. He called all his chief priests and inquired of them where the Messiah had been born. "In Bethlehem of Judea," they informed him. "See, here is what their prophet Micah has written: 'And you, Bethlehem, land of Judah, are by no means least among the princes of Judah, since from you shall come a ruler who is to shepherd my people, Israel.'"

Herod's wise men were rather perplexed. As Jews, those astrologers had come from Arabia and Ethiopia, known at that time as "The East," would have known from their sacred scriptures where the King of the Jews would be born. They seemed totally confused. They had been following an exceedingly bright star and it had sud denly disappeared. Perhaps it had been obscured by clouds? Yet it was a clear night. They were at a loss as to how such a brilliant star could suddenly disappear from view. And surely these men would have heard of Herod and would have known what an evil man he was.

They also noticed with great interest that the as trologers had asked, "Where is the newborn King of the Jews?" They did not inquire whether there was such a person. They seemed fully confident that the child born was the King of the Jews and they desired to know exactly where he was so that they might find him and worship him. If such was the case, had they never heard of Herod and that he was often and ironically called "King of the Jews?" Such a vile and vicious man would not tolerate any challenger even if that person was only a child. They also pondered the statement the astrologers had made the moment they had arrived. They had clearly indicated that they had observed "His" star at its rising and that they had come to pay "Him" homage. They wondered how such a brilliant, magnificent, and "divine" light could lead them to such an evil person like Herod. They also noticed something startling. These three men, dressed in their finest garments, had undertaken a very long and haz ardous journey bearing expensive gifts with no guards or servants to accompany or protect them. They further no ticed that if these men had come to pay homage to Herod they had not brought any gifts for him. Fearing for their own lives, they said nothing.

Herod called the astrologers aside and asked them the exact time of the star's rising. Then he sent them to Bethlehem after having instructed them: "Go and get de tailed information about the child. When you have found him, report to me so that I may go and offer him homage."

Herod showed just how cunning he was. On learn ing of the location of the child's birth he could have easily sent his soldiers into Bethlehem since it is only about six miles away from Jerusalem. He feared that armed soldiers would terrify the people, making it more difficult to find the child. Instead, he sent the astrologers on that mission.

After their audience with Herod, the astrologers set out to find the child. The star which they had observed at its rising mysteriously reappeared and traveled ahead of them until it came to a standstill over the house where the child was.

On entering the house they paid the Child homage by offering Him gifts of gold, frankincense, and myrrh. Satan did not fail to notice that and remembered that, while he was snooping around the Garden of Eden, he had observed one of the rivers there flowing through the land of Havilah, a place where there was gold and exotic spices. Satan recalled that the Queen of Sheba, journeying from Ethiopia to Jerusalem to visit King Solomon, had also come bearing gifts of gold, frankincense, and pre cious spices. Satan curiously wondered if those gifts were also from the land of Havilah or if it were just a mere co incidence. That land was where the Garden of Eden once stood and where man had fallen from grace. Those as trologers had traveled a long way from the land of Sheba —considered "The East"—and were offering those gifts to the Child, the second Adam who God had sent to redeem mankind. Satan chuckled to himself. God was sure in for a big surprise.

When the astrologers set out after conferring with Herod they were overjoyed at seeing the star again. Satan was also overjoyed. He had succeeded in leading the astrologers first to an evil king and then to the child. It was now only a matter of time before the astrologers would return to Herod, tell him of the exact location of the mother and child, and then Herod would have them slaughtered. In the battle between good and evil, Satan was positive he would emerge victorious.

But Meteron, the angel of the Lord, appeared in a dream to Joseph and commanded him to take the child and his mother and flee into Egypt; to stay there until the danger was over because Herod was searching for the child and intended to destroy him.

Meteron also appeared to the astrologers and advised them to return to their own country by a different route.

The flight into Egypt

Once Herod realized he had been double-crossed and fooled into trusting those astrologers he became furi ous. He ordered, in Bethlehem and its environs, the mas sacre of all male children who were aged, two years and under. He made his calculations based on the date of the appearance of that magnificent and exceedingly bright star the astrologers had observed at its rising. Herod was already an old man and a little child should not have caused him any worry. But Satan was fully aware of Herod's evil ways and was sure that Herod would not hesitate to get rid of any threat to his throne even if that threat came from a little child. After all, Herod had mur dered his own sons for the very same reason. To be dou bly sure that the little child born to be the "King of the Jews" did not escape his wrath, Herod ordered the mas sacre of all males under the age of two. The untold sorrow and suffering that resulted from Herod's decree brought lamentations, weeping, and great mourning that could be heard all across Judea.

King Herod soon died—a death caused by a dis ease that burned him inwardly and which was accompa nied by inexpressible torture. An intolerable stench at tended the disease and none could come near him. He was a torment to himself and a terror to all who attended him.

The angel Meteron once more appeared to Joseph and told him to take the child and his mother back to the land of Israel. Archelaus had succeeded his father Herod as King of Judea. The new king was just as ruthless as his father and Joseph was afraid to go back to Judea. He re ceived another warning from Angel Meteron and so in stead went to the region of Galilee. There they settled in the town called Nazareth.

Leslie Michael

Satan outsmarted

# 22

# Satan Outsmarted

The arrival back to Nazareth gladdened Mary's heart. This little town on the hill was her home. It was here that she grew up. It was here that the Angel Gabriel had delivered to her the most startling message—a mes sage, composed through all eternity, that she had been chosen to be the mother of Christ. It was Joseph's town as well. He, too, had grown up here, learned his trade from his father, was betrothed to Mary, and had married her. It was good to be back with family and friends. She ob served the Child growing up with great love and affec tion. She pondered all the little events in Jesus' childhood and treasured them in her heart.

Satan realized that good had triumphed over evil. All the vile and wicked schemes he had conjured up had bore no fruit. Instead of seeing the universe unfold according to his evil plans he saw it crumble before him. The world that he had envisioned and which he now found himself in was not exactly the panacea Satan had imagined. The powerful optimism and the dynamic visions that had inspired him to rebel against God had turned into perplexing doubts and anxiety. The exhilarating sense of conquest and power that Satan once felt had turned into isolation, fear, and impotence. The emptiness and the void in which he now existed made it very clear to him that his life had become worthless. The great battle for Heaven that Satan hoped would make him supreme master of the universe had turned him instead into an evil creature, cast ignominiously into a horrid place—a blazing inferno. Instead of utopia he had created hell—not only for himself but for all mankind.

Satan had attempted to create a world without spirituality and without God and felt utterly miserable to think that the angel Meteron had executed God's plan to perfection. Meteron would now ascend to Heaven to be with his God—the good, the just, the kind, the merciful and loving One. Satan envisaged Meteron in the presence of the Lord, and along with the heavenly choir of angels, singing His praises and being in perfect harmony. Satan, on the other hand, would be doomed to wander through the Earth, seeking the ruin of mankind—something he now hated because it dawned on him that he still loved God. He would give the world to be with Him, even beg for His mercy.

Surprisingly, he felt sorrow for humanity. Finding it virtually impossible to form a loving relationship with God they instead worshipped the sun, the moon, and oth er heavenly objects. Above all they craved power and ma terial goods, never realizing that one man could not over abound in external riches without another man lacking them. Stronger nations conquered weaker countries. They stole their lands and brutalized the populace. They plun dered their natural resources and made vast fortunes while most of the conquered workers lived in poverty. Even in their own countries women and children were ex ploited in factories, working long hours in appalling con ditions with unsafe machinery. Nations who wanted to re tain or regain their independence appealed to greater powers—only to be deceived as those nations had their own designs and agenda. Satan was absolutely startled when he realized that the poor, the downtrodden, the per secuted, the conquered and enslaved people hoped, prayed, and dreamed of another great war between God and Satan—one that would bring the evil, wicked, and wretched world to a blessed and merciful end.

Satan realized that while his heart was filled with hate and revenge, God had manifested Himself with love by sending His only begotten Son to redeem the world. While Satan desired power, God had sent a little infant, defenseless and without weapons. There was no intent to conquer through violence, treachery, and power. Instead God had desired that man would welcome Him in free dom and peace. God had become a defenseless innocent child to overcome man's greed, arrogance, violence, and his constant desire to possess more and more earthly goods. God had chosen the humble conditions of a manger so that man, seeking Him, would have to humble himself and stoop to enter. Satan noticed that the shep herds who had humbled themselves and entered had re turned to their homes with profound joy and gladness.

Unfortunately, he would now and forever live in a world of darkness, lies, and deceit. He would have to live eternally in a universe devoid of the divine presence of God. Sadly, and with a broken heart, Satan realized that God had not placed any chains to bind his hands. All God wanted was love that would bind his heart. As he glanced heavenward, he was aware that he was not really the op posite of God but of Michael. He recalled coming face to face with Michael in his quest to conquer Heaven and as he stood on the very edge of that abyss he had observed the kindness and compassion in the eyes of Michael. While Satan was so foolish and overwhelmed with pride, Michael was a loving and loyal angel. Brilliant as the com mander-in-chief of the angelic hosts and courageous in battle, Michael also had the intelligence and wisdom to understand God's divine plan. He was able to see God's image in man and humble enough to be the first to pay homage to Adam.

Satan was tempted to ask Angel Meteron to take a message to Michael. Perhaps Michael would intercede with God on his behalf. But fear and dread overcame him as he realized that Meteron was one of those courageous commanders who had fought valiantly and had defeated him and his rebellious angels.

Satan had made a terrible mistake and he had re paid a loving and kind God with his evil and wicked in tentions. But a mistake need never be final. Would God ever forgive him? Hope always burns eternal. As he watched angel Meteron ascend to Heaven, his heart ached and ached and Satan wept bitterly.

Leslie Michael

# Epilogue

**O**ften have I wondered why God would create Paradise and place evil in it. The Sacred Scriptures are in deed the inspired Word of God, yet, they are irritatingly elusive and appear frustratingly inconclusive, leaving the reader begging for more. We are told from the very begin ning that God created the world bringing order out of chaos. In his infinite wisdom and according to His divine will, He created man from dust, creating him in His own image and likeness. Then along came a serpent—a talking serpent who was the epitome of everything that was wicked. How then was it possible for evil to be present in the wondrous and marvelous creations of a Supreme God and allowed to mingle with man and woman, the noblest of His creations?

What is equally intriguing in the Bible is the story of three men who came to pay homage to the child, Jesus. There is no mention of who they were, where they came from, and where they departed to. The gospel of Matthew tells us that they were astrologers. If so, then they were known in ancient times as inspectors of the universe and, as such, would be naturally inclined to study a strange and exceedingly brilliant star that was moving.

We are told that these were wise men—wise be cause they followed a brilliant star. Yet these wise men did not know where the infant Jesus was born and had to ask King Herod's wise men for that information. The gospel of St. Matthew also clearly states that when they eventually found the Child he was in a house and not in a manger. Many believe they came from the East, that is, Persia or the surrounding areas. However, the Prophet Isaiah, 60:6, states, "*All from Sheba shall come bearing gold and frankincense, and proclaiming the praises of the Lord.*" The land of Sheba is in Arabia and it is not the "East." Christians also firmly believe that these men were kings. The association of Matthew 2:11 with Psalm 72:10 occasioned the mistaken supposition that the astrologers

were kings. Psalm 72:10 clearly states that it is from Ara
bia and Sheba that they shall come to pay Him homage.

The most puzzling of all is the Star of Bethlehem.
How could a divine light lead three men looking for the
child that would be the King of the Jews; lead them to the
wicked King Herod, knowing fully well that Herod
would not tolerate any threat to the throne? How did a
brilliant and exceedingly bright star disappear from view
and then reappear to guide the three men to Christ—after
leading the men *first* to Herod?

The church seemed very uncomfortable with as
trology and therefore chose to ignore the power, the re
sourcefulness, the cunning and evil ways of Satan and
hastily proclaimed the star to be a divine and miraculous
event.

My task was to write this book clearly, logically,
and with no intention of offending anyone. It is, in the fi
nal analysis, the story of the greatest game of chess that
was played between God and Satan. In the beginning
when God made it clear to the angelic hosts that He
would create a new world and would place in it a creature
that would be in His own image and likeness, Lucifer did
not agree with God's plan. His pride prevented him from
perceiving God's divine plan and he rebelled because he
refused to pay homage to man who would be created

from dust and who would be far inferior to him. God's first move was to have him cast out of Heaven and a great battle ensued between Lucifer and Archangel Michael, the great commander of the heavenly host. Lucifer, now Sa tan, lost the battle for Heaven and was cast into a sea of flames.

Satan was not about to take this lying down. He re membered God telling the angels about a new world and a creature called "man." He set out on a quest to find this person and was pleased when he found him in the Gar den of Eden. He cunningly and craftily entered the Gar den of Eden as a serpent and was successful in bringing about the fall of man. Adam and Eve lost their immortali ty and were driven out of Paradise.

It was now up to God to make a move. Satan's tri umph was short-lived. In His infinite love and mercy, God sent His only begotten Son to redeem mankind and give man another chance at eternal life.

Satan, infinitely cunning, deceptive, and powerful was livid with rage. He remembered that he was one of the most brilliant angels in Heaven, created from fire, em bedded with precious stones, and brighter than the noon day sun. He was endowed with great intellectual and physical powers and capable of appearing in any shape or form. It was Satan's move now and he used those im mense powers. He appeared as a brilliant and magnificent star and guided three astrologers to Herod, knowing fully well that the vile and wicked Herod would not tolerate any threat to his throne even if that threat came from a lit tle child.

When it was God's turn, He sent the Angel Me teron to warn the Holy Family of Herod's murderous in tentions.

This story tells us of God's triumph and Satan's de feat. It is the story of a magnificent angel who was cast out of Heaven; a rebellious angel who swore vengeance; a revengeful angel who dared to challenge God.

Good had triumphed over evil and Satan realized that he was lost, lonely, and in a terrible place, totally alienated from God. He was heartbroken because he real ized that he still loved God and wondered if God would ever forgive him.

The book is written in a non-confrontational man ner. It is not meant to offend anyone. My fervent hope is that this book will open the door to an intelligent dialogue and meaningful discussions.

Leslie Michael

Coquitlam, British Columbia,

Canada

March 15, 2010

# Bibliography

1.  *The Holy Bible, with Illustrations by Gustave Doré.* London & New York: Cassell, Petter and Galpin, 1866-70.

2.  *Paradise Lost,* John Milton, illustrated by Gustave Dore. New York. Peter Fenelon Collier, 1866

3.  *Paradise Lost.* John Milton. An Authoritative Text. Backgrounds and Sources Criticism. A Norton Critical Edition, edited by Scott Elledge. Cornell University. Sec ond Edition W.W. Norton & Company. New York. Lon don. Library of Congress Cataloging-in-Publication Data: Milton, John 1608-1674. ISBN 0-393-96293-8.

4.  *The New American Bible.* Translated from the Original Languages with Critical Use of All the Ancient Sources by Members of the Catholic Biblical Association of America. Sponsored by the Bishops' Committee of the Confraternity of Christian Doctrine. Catholic Book Pub lishing Co. New York The text of the NEW AMERICAN BIBLE, copyright @ 1970 by the Confraternity of Chris tian Doctrine, Washington, D.C.

5.  *Matthew Henry's Commentary On The Whole Bible with Practical Remarks and Observations.* Volume 1 – Genesis to Deuteronomy. New York. Fleming H. Revell & Company. London and Edinburgh.

6.  *Kebra Nagast.* The Queen of Sheba and her only son, Menyelek, being the 'Book of the Glory of Kings' (Kebra Nagast). A work which is alike the traditional history of the establishment of the religion of the Hebrews in Ethiopia and the patent of sovereignty which is now universally accepted in Abyssinia as the symbol of the Divine Authority to rule which the Kings of the Solomonic line claimed to have received through their descent from the House of David. Translated from the

Ethiopic by Sir E. A. Wallis Budge. M.A.,LITT.D., D.LITT., LIT.D., F.S.A Sometime Scholar of Christ's Col lege, Cambridge. Tyrwhitt Hebrew Scholar and Keeper of the Department of Egyptian and Assyrian Antiquities in the British Museum. MCMXXXII. Oxford University Press, London. Humphrey Milford. February,1922.

7.  *The Twelve Caesars* by Gaius Suetonius Tranquillus. Translated by Robert Graves. Revised with an introduc tion by Michael Grant. Penguin Classsics. Published 1957. Printed in Great Britain by Hazell Watson & Viney Ltd, Founder Editor (1944-64) E.V. Rieu. Editor: Betty Radice .

8.  *Patriarchs and Prophets.* How It All Began. E. G. White. Printed for Remnant Publications by Pacific Press Pub lishing Association, Boise, Idaho, USA and Oshawa, On tario, Canada. Published 1993. ISBN 1-883012-50-3

9.  *Two From Galilee.* A Bantam Book by Marjorie Holmes. Published by arrangement with Fleming H. Revell Company. May 1972. ISBN 0-553-14770-6

10. *All the Women of the Bible.* Edith Deen. Harper-Collins Publishers Inc. 10 East 53$^{rd}$ Street, New York, NY. 10022. Library of Congress Catalog Card Number 55-8521, ISBN 0-7858-0471-4.

11. *Cedar Of Lebanon*, John Cosgrove. 1952 McMullen Books, Inc., 22 Park Place, New York 7, N.Y. USA.

12. *A tale of Two Trees.* Serpents of Desire. Part 2. Rabbi David Fohrman. Jewish History Crash Course. Rabbi Ken Spiro .

13. *The Battle For God.* Karen Armstrong. A Borozoi Book. Published by Alfred A. Knopf. New York. 2000. ISBN 0-679-43597-2

14. *The Screwtape Letters.* Revised Edition. C.S. Lewis. Collier Books. Macmillan Publishing Company. New

York. 1982. ISBN 0-02-086740-9. The definite edition containing C.S. Lewis Preface of 1960, Screwtape Pro poses a Toast and the never-before published Preface to the Toast.

15. *American Caesar.* William Manchester. Little Brown & Co. Ltd. Published September 2008. ISBN 0091365104 .

16. *The Source.* James Michener. Random House. USA 1965 .

17. *The Journey of the Magi.* Paul William Roberts.

18. *The Russian War machine 1917 – 1945.* Edited by S.L. Mayer. Chartwell Books Inc. Distributed by Book Sales Inc., Secaucus, New Jersey, USA.. Copyright 1977 by Bi son Books, London, England. Printed in Japan ISBN 0-89009-082-3. Library of Congress Catalog Card Number: 77 – 71724. The Russian War Machine was the most for midable fighting weapon of World War II. More than any other single factor, the Soviet Army and Air Force destroyed Nazi power and brought victory over Ger many and Europe.

19. The Internet. Various articles and references from Wikipedia.

20. The Encyclopedia Britannica